Taking Appearance Seriously

'*Taking Appearance Seriously* is a rare philosophical work of both outstanding quality and immense practicality, written to guide the reader into really experiencing what Henri Bortoft calls the dynamic way of seeing: a radically aware way of thinking and comprehending our complex world which is as applicable in the creative arts and business world as it is in science.'

SIMON ROBINSON, EDITOR OF *TRANSITION CONSCIOUSNESS*

—

'Bortoft's aim is to help readers see and understand the world and human experience in a more integrated, compelling way. Exploring and developing previous work on the fascinating relationship between parts and whole through the lenses of phenomenology, hermeneutics, and Goethean science, Bortoft demonstrates convincingly the need to shift attention from what is experienced to the experience of what is experienced. In a writing style that is both accessible and penetrating, he shows how to do this by drawing on examples from our experience of meaning, understanding and language. By teaching ourselves to become more sensitive to this dynamic way of seeing, we learn to "take appearance seriously".'

DAVID SEAMON, PROFESSOR, KANSAS STATE UNIVERSITY, AND EDITOR, *ENVIRONMENTAL AND ARCHITECTURAL PHENOMENOLOGY*

—

Henri Bortoft is one of the world's foremost experts on Goethean science. Here he extends the argument of his earlier work The Wholeness of Nature to articulate a new history and philosophy of science with an emphasis on a dynamic way of understanding that highlights process rather than product, the coming into being rather than the end result. This represents a fundamental shift away from abstraction to lived sensory experience, and from the dominance of left-hemisphere thinking to a more integrated approach. This is a seminal text that deserves the widest audience.

DAVID LORIMER, PROGRAMME DIRECTOR OF THE SCIENTIFIC MEDICAL NETWORK

Taking Appearance Seriously

Seriously

The Dynamic Way of Seeing
in Goethe and European Thought

HENRI BORTOFT

Floris Books

First published by Floris Books in 2012
© 2012 Henri Bortoft

This book is also available as
an eBook

British Library CIP Data available
ISBN 978-086315-927-5
Printed in Great Britain
by TJ International Ltd, Cornwall

For Ella, Ben and Darcy

Contents

Preface and Acknowledgments

This book has taken longer to write than expected. Although the idea for it was conceived about ten years ago, the roots of it go back much further. My wife, Jackie, has been unfailingly patient and supportive throughout, putting up with a husband who was obsessed by an idea and yet unable to say clearly what it was. She has typed the many drafts I have written, pencil in one hand and rubber in the other, and also drawn the diagrams. I owe her a great debt for her help and loyalty throughout what must at times have seemed interminable, and without which it is true to say this book could not have been written.

I also want to thank Christopher Moore, my editor at Floris Books. His enthusiastic response encouraged me to bring the book into its final shape, and I believe his suggestions will make it more accessible to the reader. It has been a stimulating experience for me to find my work being read by an editor who grasps what it is about so clearly.

The first three chapters have benefited from working with students on the MSc Holistic Science course at Schumacher College, where for several years I have had the privilege of teaching a module on the philosophy of holistic science. This opportunity – and especially the quality of attention the students have given – has helped me to understand Goethe's way of seeing better than I would have done otherwise. Chapters 2 and 3 can stand on their own for anyone who is solely interested in Goethe. The fourth chapter benefited from being given as a workshop on hermeneutics in New York in 2008, and I am very grateful to Gary Gomer for suggesting this and making it possible.

Finally, I want to say that this book is more 'practical' than it looks. Above all, it is not nearly as difficult to follow as the reader unfamiliar with European philosophy might expect. I have tried to write it in such a way that anyone who reads it slowly enough to follow the movement of thinking in the language, should find they begin to experience the dynamic way of seeing for themselves.

1. Into the Dynamic Way of Thinking

Philosopher consiste à invertir la direction habituelle
du travail de la pensée

HENRI BERGSON

This is a book about a different way of thinking. The dynamic way of thinking – which is the general name I am going to give it – first appears in European thought at the beginning of the nineteenth century with Johann Wolfgang von Goethe and the early Romantics, and the philosophers Schelling and Hegel, all of whom were in and around Jena at the same time. Here, as always, it takes a form that is specific to the particular circumstances in which it appears. Confusing the container with the content, as we so often do, means that inevitably we end up focusing too much attention on the specific form which this way of thinking takes in a particular instance, and consequently fail to see the more universal content which is the movement of thinking itself.

The dynamic way of thinking appears again in European thought in the first part of the twentieth century in the philosophy of phenomenology and hermeneutics. Here once again we are too easily seduced by the specifics of the occasion to notice the more universal element. Divergent as these philosophical movements may seem outwardly – and they *are* divergent – they nevertheless belong together when they are seen in terms of the movement of thinking which each expresses in its own different way. The significance of this dynamic way of understanding easily gets lost in the obfuscations of philosophers who, in their endless attempts to justify what they are doing, all too often succeed only in covering it over with a dense layer of what to others seems to be just impenetrable jargon. The vision gets lost, and what is left descends into an intellectual exercise, which turns round upon itself endlessly until it ceases to be of interest to any but a few. This is such a pity, because there is something here which is potentially of much wider interest and which needs to be brought out. I believe this can be done by taking a more concrete approach. This is what I am going to do here, and for this reason I am going to begin by going back to my own first encounters with the dynamic way of thinking.

My introduction to European philosophy came through an unusual route. I had been working in a small research group investigating more effective ways of communicating ideas in education. At the time – the late 1960s and early 1970s – there was a growing interest in the UK in management education and organisational development. The kind of methods for more effective communication which we were researching turned out to be also of interest here – in fact more so than in mainstream education, where institutional constraints sometimes made innovation difficult. This was at the time when 'Systems Theory' was very much in vogue in the world of management and organisation. Diagrams were much in evidence, usually consisting of words in boxes joined together by lines to represent connections. The aim of systems thinking was to move away from the emphasis on the idea of basic building blocks towards the idea of the overall order of the organisational form.

Systems thinking is often presented as a revolution in thinking that overcomes the limitations of the Cartesian paradigm of analytical thinking that has been central to modern thought. In some ways this is undoubtedly true – in the Cartesian paradigm the behaviour of the whole can be reduced to the behaviour of the parts, for example, whereas the very opposite is the case in systems thinking. However, in another respect systems thinking has a surprising affinity with Descartes' methodological goal, so much so in fact that it could even be called the ultimate fulfilment of Descartes' dream. The failure to recognise this is a consequence of selecting only part of Descartes' work for attention, instead of seeing it more comprehensively. What was central for Descartes was his dream of a *mathesis universalis* (universal mathematics), which would be in effect a seventeenth century 'unified science' or 'theory of everything'. Having shown that problems in geometry could be expressed as problems in algebra, so that figures could be eliminated from geometry, thereby unifying what until then had been thought to be two different sciences (this is what Descartes called them), he believed that it must be possible to go further in the direction of unification by eliminating quantity itself from mathematics. The resulting universal science could then apply to any subject matter whatsoever. In his *Rules for the Direction of the Mind*, he says:

> I came to see that the exclusive concern of mathematics is with questions of order and measure and that it is irrelevant whether the measure in question involves numbers, shapes, stars, sounds, or any other object whatsoever. This made me realise that there must be a general science which explains all the points that can be raised concerning order and measure irrespective of the subject-matter, and that science should be termed *mathesis universalis* [universal mathematics].

This dream of a unified science emerged again in the 1920s, some three hundred years later, among the philosophers and scientists who were part of what came to be known as the Vienna Circle. Some of these – notably Rudolf Carnap – believed that all the different sciences (including psychology and sociology) could be unified by effectively reducing all the sciences to physics, since this is the science closest to pure mathematics. Although this suggestion may seem very strange to us today, this gross reductionism was embraced enthusiastically by some until the 1960s. However, another member of the Vienna Circle, Ludwig von Bertalanffy, advocated a different approach which led eventually to what he called General Systems Theory. Instead of producing unification by reducing all sciences ultimately to the method of physics, von Bertalanffy proposed a mathematical science of general systems which would apply to all systems irrespective of their nature, whether they be physical, chemical, organic, ecological, psychological, sociological, cultural or historical. He said that, just as the mathematical theory of probability deals with 'chance events' as such, irrespective of their nature, so general systems theory would deal with 'organised wholes' as such. It would apply to all the sciences – physical, biological, psychological, sociological, and even to history. As he put it, the 'Unity of Science is granted, not by a utopian reduction of all sciences to physics and chemistry, but by the structural uniformities of the different levels of reality'.[1] If we compare this with the statement made by Descartes concerning the idea of a *mathesis universalis,* even allowing for the differences between them as a consequence of their being three hundred years apart, von Bertalanffy's science of 'the structural uniformities of the different levels of reality' sounds very similar to Descartes' 'general science of order and measure irrespective of the subject matter'. It seems that, unbeknown to him, von Bertalanffy was pursuing the same ideal that was first introduced into modern western thinking by Descartes. This is ironic, because there are many today who believe that it was systems thinking which first overcame the reductionism so often associated with the name of Descartes. It seemed to me that, although the claim was made that systems thinking is holistic, and therefore non-reductionist, it is in fact much more reductionist in practice than many of the optimistic pronouncements about it would lead us to suppose.

A Different Approach to Wholeness

My main concern was with the claim that systems theory is a science of wholeness. This arose out of my experience as a postgraduate research student in physics at Birkbeck College early in the 1960s, where I worked on the problem of wholeness in the quantum theory. It had become clear that a fundamentally new way of thinking was needed for quantum physics, even though such a possibility had been explicitly denied by Niels Bohr in what was referred to as the Copenhagen interpretation of the quantum theory, which had become the most widely accepted view among physicists as a result of Bohr's extraordinary persuasiveness. But David Bohm believed this could be done. He pointed to examples which he said could function as templates for a new way of thinking about wholeness. One of these was the hologram – which at the time in question was a technological innovation. This appealed to the imagination because, unlike a photographic plate (where each point of the image on the plate corresponds approximately to a point on the object), with the hologram each part of the plate contains information about the whole object. Thus instead of localised parts, with the hologram the whole is present in each part and each part is distributed throughout the whole. To use the language which Bohm later adopted: the whole is enfolded in the part and each part is enfolded in the whole.[2] These ideas of Bohm's encouraged some of us to think that the wholeness of human organisations, at whatever level, could not be understood adequately by means of the systems approach because something more 'holographic' was needed.

One of the areas in which we were working required the design of an 'attitude survey' for the preliminary stage of gathering information prior to the introduction of an organisational change. We adopted the philosophy that each person in their role in an organisation is in fact an expression of the organisation as a whole, so that we could say the whole organisation comes to expression, to some degree, through the role of each person in that organisation. So if the whole comes to expression through its parts – which will therefore each include reflections of all the others to some degree (i.e. they are internally related) – then the way to understand the whole is through the way it is expressed within the parts, instead of trying to stand back to get an overview to see how the parts could be made to fit together into a whole – which all too often seemed to be the outcome, if not the intention, of the systems approach. In practical terms, if the way into the whole is through the parts, each of which is an expression of the whole, instead of trying to get a total overview of the whole, then this meant talking to everyone in the organisation because, whoever they were, the whole was coming into expression through them, no matter how partially. Encountering the whole in this way felt like entering into another dimension of the organisation – but a concrete dimension – compared with the usual way of thinking. Our practical task, as we interpreted it, was to devise surveys and other materials which would facilitate this 'holographic' approach to the wholeness of the organisation in which we could begin to see the wholeness from within the organisation, instead of trying to 'see it as a whole' by standing outside of it.

One day I was trying to describe the idea behind this work to Brian Lewis, who was professor of educational systems at the Open University. He told me that it sounded to him very similar to what is called 'the hermeneutic circle', and he suggested that I looked into the philosophy of hermeneutics.[3] This philosophy arose in the first place in connection with questions about how we understand written works – whether they be scriptural, philosophical, literary, historical, or legal. But it became apparent that hermeneutics applies more widely to all forms of expression, and hence to any kind of cultural expression from the simplest to the most complex. Put simply, if somewhat abstractly, the hermeneutic circle arises from the circumstance that, in order to understand the whole we must understand the parts, but in order to understand the parts we must understand the whole. It became obvious immediately that the holographic approach to wholeness – with which it was intended to replace the systems approach – had a form which is very similar to that of the hermeneutic circle, and hence that what we thought of as a 'holographic' survey could equally well be thought of as a 'hermeneutic' survey. Switching from the holographic model to hermeneutics, had the advantage that it located what we were trying to do in the context of a known, even if unfamiliar, philosophical tradition. This opened the door to the possibility that systems thinking could be replaced by hermeneutic thinking in the context of human organisations. There was an explosion of activity as some of us explored the hermeneutic dimension of the organisation in as many ways as we could find – which included one occasion when I found myself giving a seminar on 'The Hermeneutics of the Organisation' to the somewhat bemused management of IBM.

I tried to express the difference between this and the systems approach in a paper which I gave at a conference at the beginning of the 1970s.[4] What I wanted to do in this paper was to find a way of talking about wholeness that would avoid the 'totalitarian' tendency of systems theory – as a result of which the whole is *reified* and separated from the parts which it then dominates. The aim is to avoid reductionism without replacing it by holism. The hermeneutic circle gives us a different way of thinking, in which the parts depend upon the whole, but equally the whole depends on the parts. I found the language I was looking for in Heidegger's notion of 'presence' (not to be confused with 'present'), 'presencing', 'coming-to-presence', and so on. This enabled me to say that the whole *presences* within the parts, which is intended to convey the sense that it is always implicit and can never become explicit as such – if it did it would become 'present' as an object (it would come 'outside') and hence separate from the parts. If the whole presences within the parts, then the only way to encounter the whole is within the parts through which it presences, and not by standing back from the parts to try and get an 'overview' of the whole. In her *Safeguarding Our Common Future,* Ingrid Stefanovic gives a beautiful illustration of this:

> At the very least a new way *of seeing* things seems to be called
> for. I am reminded of my first experiences in photography,
> when I lived in a particularly beautiful section of Victoria,
> British Columbia some years ago. The spectacular houses

and gardens of Oak Bay had been part of my everyday world for only a few months when I resolved one weekend to meander through my neighbourhood, capturing images through the lens of my new camera. For the first time, I took note of details of leaded windows, garden fountains and pools, and flowers that were, miraculously, already blooming in February.

The experience led me to realise that, while the camera focused my attention on specific aspects of my neighbourhood, what made these images special was that they constituted more than an isolated, atomistic parcelling up of the neighbourhood through the camera lens. Instead, each image was significant inasmuch as it captured and articulated in a distinctive way, the sense of place of the neighbourhood as a whole. On the one hand I was drawn to notice particular details that I had missed, when I had not sought them out through the lens of my camera. On the other hand, each individual photograph was all the more meaningful to the degree that the broader sense of the place as a whole was reflected and even in some sense enriched in each photographic image.[5]

This 'resonance of the whole sense of place within the perspective of each individual photograph' is clearly an instance of the coming-to-presence of the whole within the parts.

Looking back now, it seems to me that the difference between the two approaches to wholeness reflects the difference between the world as mediated through the two hemispheres of the brain. Although the experience of wholeness has always been identified with the right half of the brain, it is now recognised that *every* characteristic of experience is in fact mediated through *both* sides of the brain, and consequently this must also be the case with wholeness. According to Iain McGilchrist: 'the right hemisphere delivers what is new as it "presences" – before the left hemisphere gets to represent it'.[6] Where the right hemisphere mediates the lived experience of wholeness, the left hemisphere mediates its representation – it replaces experience with a model of experience, which then gets confused with and mistaken for experience itself. The wholeness of the system is the left brain representation of the wholeness which presences through the right brain. This explains why it is that the systems approach *seems* to be dealing with wholeness, but does so in an artificial way that is a counterfeit of authentic wholeness.

Introduction to Phenomenology

This interest in hermeneutics led me quite naturally into phenomenology – the most important and influential movement in European philosophy in the twentieth century. Hermeneutics as the philosophy of meaning and understanding was transformed by phenomenology, first through Martin Heidegger and then by Hans-Georg Gadamer. But getting into phenomenology isn't easy. It is a philosophy which has the effect of seeming strange and yet familiar at the same time. Phenomenology seems to take the ground away from under our feet, whilst at the same time giving us the sense of being where we have always been – only now recognising it as if for the first time. It's hard to catch hold of because it's like trying to catch something as it's happening and which is over before we can do so. It can perhaps be described most simply as 'stepping back' into where we are already. This means shifting the focus of attention *within experience* away from what is experienced into the experiencing of it. So if we consider seeing, for example, this means that we have to 'step back' from *what* is seen into the *seeing* of what is seen. Like many others, I felt drawn towards phenomenology and yet frustrated by it, because it seemed to be both evident and elusive at the same time. However, by good fortune the stirring of my interest in phenomenology happened to coincide with the founding of the British Society for Phenomenology by Wolfe Mays. This gave me the opportunity to meet and learn from practitioners – which included not only academic philosophers, but also psychiatrists, sociologists, and others, who used the phenomenological approach in their work. It was like breathing in an atmosphere of phenomenology, and under these circumstances it wasn't long before I began to 'catch' the phenomenological way of seeing.

It was against this background that I was asked to give a series of workshops on phenomenology and hermeneutics at a new residential adult education centre.[7] The aim was not to fill the students' heads – and notebooks – with intellectual material on what this or that philosopher said, but to bring them to the point where, some of them at least, could begin to get a taste of this way of seeing for themselves. It seemed like an excellent opportunity. However there were a number of drawbacks, not the least of which was the fact that I hadn't the faintest idea how to do it. But, overcome by the enthusiasm of youth, it became a case of the proverbial fool rushing in where angels would fear to tread. I simply hoped that I would get a clearer idea of how to proceed with these workshops the nearer it got to the time. But this didn't happen, and my anxiety level began to rise the closer it got – especially when I learned that I would be expected to take three different groups of adult students, each for two sessions a week, for a total of twelve weeks (thankfully with a break after the first six weeks). When I took up residence at the college the day before I was due to begin, I went for a walk in the countryside in the hope that this might at least have the effect of reducing the level of anxiety I was now experiencing. I made my way to the bottom of the valley through which a small, clear river ran. I stood on a bridge, looking downstream at the river flowing away from me. For some reason this made me feel uneasy, and I crossed to the other side to look at the river flowing towards me. This felt better, and I spent some time there, looking upstream. I began to be drawn into the experience of looking, plunging with my eyes into the water flowing towards me. When I closed my eyes I sensed the river streaming through me, and when I opened them again, I found that I was experiencing the river flowing towards me outwardly and through me inwardly at the same time. The more I did this, the more relaxed and free from anxiety I began to feel. But of course, the moment eventually came for the first workshop to begin. I remember walking down the long corridor toward the room where it was to take place, feeling I was about to be extinguished. The door at the end was closed, the students were already waiting inside, and as I turned the doorknob to go in I expected to fall into an abyss on the other side. Instead, as I walked into the room, I heard myself saying, with surprising confidence: 'Our problem is that where we begin is already downstream, and in our attempt to understand where we are we only go further downstream. What we have to do instead is learn how to go back upstream and flow down to where we are already, so that we can recognise this as not the beginning but the end. That's phenomenology!' I don't know who was more surprised, myself or the students. It was a good start, a doorway into the movement of thinking in phenomenology through which after that I found I could begin to go.

intrinsic direction of experience

———————————————————————▶

{the *experiencing* of what is experienced} ◀——— *what* is experienced

The Act of Distinction

Phenomenology is a shift of attention within experience, which draws attention back from *what* is experienced – i.e. where the focus of attention is on the *what* – into the *experiencing* of what is experienced: The { } is important. If we just say there is a shift of attention from *what* is experienced to the *experience*, we are in danger of unwittingly treating 'experience' as if it could be separated from what is experienced. But there can be no experience without something that is experienced. The shift of attention 'back upstream' is subtle, and not coarse as it would be if we made the mistake of trying to focus on 'experience' directly – this would mean trying to turn experience into *what* is experienced, which is the fallacy of introspection with which phenomenology has often been confused.

So, for example, if we are concerned with seeing, the 'phenomenological move' is to shift the position of attention within experience back from *what* is seen into the *seeing* of what is seen:

{the *seeing* of what is seen} ◀————————————— *what* is seen

If we are concerned with saying, we have to draw attention back from *what* is said into the *saying* of what is said:

{the *saying* of what is said} ◀————————————— *what* is said

When we do this we discover that the 'common sense' account of perception (empiricism) and language (nominalism) are not true to experience.[8] But the example we are going to consider first is the act of

19

distinction, in which case we have to draw attention back from what is distinguished into the *distinguishing* of what is distinguished:

{the *distinguishing* of what is distinguished} ◀—— *what* is distinguished

When Goethe read a translation of Luke Howard's seminal essay *On the Modification of Clouds*, he said that Howard was 'the man who distinguished cloud from cloud', and he wrote a poem in his honour in which he said Howard had 'Defin'd the doubtful, fix'd its limit-line, and named it fitly'. It may seem extraordinary to us today that Howard's simple classification of cloud formations – cirrus, cumulus, stratus – could be the source of so much scientific excitement and widespread admiration. At the time it was quickly recognised that Howard had opened the door (which others had also sought and failed to find) to the scientific study of meteorology, but now we would look upon this as if he had done no more than impose a system of classification simply by applying labels externally to the superficial appearances of the clouds. But this is because we begin 'downstream' with the end result, the system of names, instead of going 'upstream' into the process of discovery to glimpse the coming-into-being of the distinction of which these names are the expression.

How could anyone find a natural order in the ever-changing phenomena of the clouds? The very idea of finding anything fixed and constant in such fluid and impermanent phenomena seems at first absurd. Yet Howard was able to discern the hidden dynamics of the clouds, and thereby distinguish three fundamental cloud types which he said are 'as distinguishable from each other as a tree from a hill, or the latter from a lake'.[9] He was able to show that the teeming myriads of cloud formations are all modifications of only *three* types (where we might have expected to find a multitude, or even none at all) forming and transforming into one another according to the atmospheric conditions. As Goethe and others recognised, Howard distinguished the cloud formations, not in the sense of classifying them according to secondary characteristics, but in a unitary act of {differencing/relating} in which the types are seen as simultaneously different from and related to one another (this will be discussed below). We could say that, in both senses, Howard articulated the clouds, because distinguishing and naming are two sides of the same coin. This example shows clearly that the act of distinction is simultaneously analytic and holistic. Although when we begin at the end it seems to result in no more than a division into separate categories – difference 'falls apart' into separation – when we try to catch distinction 'in the act' we find that it is not divisive but holistic. Thus, when he 'distinguished cloud from cloud', Howard simultaneously revealed the dynamic wholeness of the phenomenon – as Goethe clearly recognised.

Heidegger's distinction between *belonging* together and belonging *together* is helpful here.[10] In the first case the belonging is primary and determines the together, whereas is the second case it is the together which determines the belonging. Thus, in the latter case, we bring things together, or put them together, and say that now they belong with one another because we have togethered them. But in the case of *belonging* together it is the other way round. Here things already belong with one another and this belongingness determines their togetherness. We can now begin to appreciate the difference. *Belonging* together is subtle, and if we do not become aware of the way in which things already belong, then we may try to make them belong by togethering them – i.e. by imposing a framework which organises them. Since this will not be sensitive to the more subtle way in which things already *belong* together, the organisational framework that brings them together can only be imposed externally and not be intrinsic. Hence it is coarse. The wholeness of the system is basically that of a framework which organises by togethering, and which all too often eclipses the more subtle wholeness of belongingness. In terms of Heidegger's distinction, we could say that Howard revealed the *belonging* together of the clouds, instead of trying to make them belong *together* by imposing an external system of classification. This is the distinction between authentic and counterfeit wholeness.

When we think of the act of distinction in terms of the outcome – i.e. in terms of *what* is distinguished – we cannot avoid thinking of distinction *only* in terms of difference – that one thing is different from another – and the movement of thinking here is one which almost automatically turns distinction into separation. So we come to think that 'distinction' and 'separation' are the same. But they are not. We can see that they are not the same by trying to go 'upstream' into the *act* of distinction itself – which means going into the *happening*, the coming-into-being, which is the *appearance* of distinction. We could call this dynamical distinction the primary distinction, as opposed to the secondary distinction which merely partitions and separates what has already been distinguished. When we go 'upstream' and try to 'catch distinction in the act', we discover something fundamental which we overlook when we begin 'downstream' with *what* is distinguished. When we shift our attention into the happening which is the appearing of distinction, we notice that distinction not only 'differences', but that at the very same time it also relates.[11] It is when we focus *only* on the difference – as we do when our attention is focused on *what* is distinguished, the outcome, instead of the act itself – that we confuse distinction with separation.

We say that A is distinguished *from* B, or that X is distinguished *from* its surrounding (which thereby become the background against which X stands out as being X). We must remember here that we are describing the very *act* of distinction, and so we must not fall into the trap of thinking of A and B, or of X and its surroundings, as if they were already there as such, so that the 'distinction' would amount to no more than separating what is already distinguished – in which case we are already 'too late' in our thinking to catch the distinction 'in the act'. If A is distinguished from B, or X from not-X, then the very act of distinction which differences simultaneously relates – i.e. if A is distinguished *from* B, it is thereby concomitantly related *to* B by the very act which distinguishes it. Since this relation is *intrinsic* to the distinction, and not added afterwards, it is called an 'internal relation'. It is as if the act of distinction goes in opposite directions simultaneously. Distinguishing is a dual movement of thinking which goes in opposite directions at once: in one direction it differences, whereas in the other direction it relates. So the *act* of distinction 'differences/relates' – not differences *and* relates, because this would be two movements, whereas there is *one* movement which is dual. What comes into being as a distinction is therefore a difference/relation and the act of distinction is a *unitary* act which {differences/relates}. If the relation which is intrinsic to the distinction is not noticed, then the distinction can only turn into separation – which is what happens when our attention shifts from the *distinguishing* of what is distinguished to focus on *what* is distinguished. When this happens, so that distinction is thought of only in terms of separation, it seems that the act of distinction is just analytical. But when we follow the coming-into-being of distinction we recognise that it must also be holistic. This is not something we would have expected to find.

It may be helpful to find an image for this simultaneity of what seem to be opposites, i.e. difference/relation and analytic/holistic. The biperspectival figure which is familiar from gestalt psychology may be useful here – the duck/rabbit for example (see Figure 1).

Figure 1. The ambiguous duck/rabbit image.

As this is not duck *and* rabbit, but simultaneously duck/rabbit, so the act of distinction is simultaneously the analytic/holistic act of {differencing/relating}. By reflecting on such an auxiliary, we can see how it is not a case of partly one and partly the other, but of one which is simultaneously both.

The *happening* of distinction is the *appearing* of what is distinguished. It is well-known that when something is first distinguished it soon appears to all who are able to see it, whereas previously it had not been seen by anyone, even though once it has been distinguished we feel it was there to be seen all along and we are astonished that nobody actually did see it. The medical disorder of muscular dystrophy provides an illustration of this. Before the 1850s, when this disease was first described (that is, distinguished) by the French neurologist Guillaume Duchenne, it had not been recognised by anyone. But, once distinguished, what had not been seen before began to be widely recognised, and by the 1860s many hundreds of cases had been seen and described. This prompted his contemporary Jean-Martin Charcot to comment: 'How come that a disease so common, so widespread, and so recognisable at a glance – a disease which has always existed – how come that it is only recognised now? Why did we need M. Duchenne to open our eyes?'[12] Being able to recognise it depends on the primary act of distinguishing muscular dystrophy, so that it stands out.

What we later consider to have been there in front of us all the time is invisible to us before it is distinguished – we could say that the act of distinction 'theres' it.

What we come to here is something remarkable: the appear*ance* of 'muscular dystrophy'. There is a shift 'upstream' here: 'when we speak of what 'appears', we refer not only to a *thing* but to a *happening*: the appearing itself'.[13] There cannot be appearing without something that appears, but we can shift the focus of attention *within* experience from *what* appears into the *appearing* of what appears:

{the *appearing* of what appears} ◄——————————— *what* appears

This is the fundamental phenomenological step – the examples of seeing and saying, as well as distinguishing, are all really specific instances of appearing. In a lecture given in 1907, Edmund Husserl points out that:

> The word 'phenomenon' is ambivalent because of the
> essential correlation between appearance and the appearing.
> According to this notion a phenomenon is not only
> something which appears, but something which appears *as*
> *appearing.*[14]

The crucial point is that phenomenology is concerned with what appears *in its appearing.* So the phenomenon is not merely the appearance but the appear*ance.* This is the phenomenon: the appearing of what appears. If we don't understand this, and instead think that the phenomenon is merely the appearance, then we miss what phenomenology is really about and can easily confuse it with phenomenalism.

We cannot describe Duchenne's discovery of muscular dystrophy epistemologically, in terms of a subject knowing an object, because in this case the object itself only appears in being known. The epistemological framework is already too late. But this does not mean that the discovery is simply subjective. Duchenne didn't just find muscular dystrophy, but then neither did he produce it. We have to find a way of thinking which 'splits the difference between "finding" and "making"'.[15] Clearly this is paradoxical to our either/or way of thinking. What we are looking for here is expressed very clearly by McGilchrist:

> One way of putting this is to say that we neither discover an objective reality nor invent a subjective reality, but that there is a process of responsive evocation, the world 'calling forth' something in me that in turn 'calls forth' something in the world.[16]

So the dynamics of appearance is that something in the world [which has not appeared] evokes a response [in the perceiver] which calls forth that in the world which evokes this response [it appears]. It is a dynamical whole – but the reciprocity is asymmetrical. In the language of Husserl's *Fundierung* relation, the found*ing* term has an originality or priority in that the found*ed* term is derived from it, but as Merleau-Ponty points out, it is not 'simply derived', because it is through the founded term that the founding term manifests – 'it is through the originated that the originator is made manifest'.[17]

It looks like we create what at the same time we seem to discover, and this seems paradoxical. But McGilchrist points to an earlier tradition in the history of philosophy (which Heidegger has retrieved) for which 'the act of creation may be ... one of discovery, of finding something that was there, but required liberation into being'.[18] In such a case, where discovery means freeing the entity into appear*ance*, we are 'finding something which is coming into being through our knowing, at the same time that our knowing depends on its coming into being'.[19] 'Coming into being' here means 'appearing'. This is why Heidegger says:

> Being means appearing. Appearing is not something subsequent that sometimes happens to being. Being presences *as* appearing.[20]

This is astonishing – and very easily misunderstood.[21] It removes the *separation* between being and appearance which is so familiar in the metaphysical tradition. There is no longer the dichotomy of being and appearance which is the ultimate dualism, and 'the curse of mereness' is lifted from appearance.[22]

But this is only possible if we go 'upstream' from appearance to appear*ance*, when we will see that there is no separation because being *is* appear*ance*. If we don't, and think that 'appearance' just means what appears, the look of things, then we will miss the dynamics of being and think that Heidegger just reduces being to appearance. Phenomenology liberates us from the dualism of metaphysics, but without leading us into phenomenalism – which is usually seen as the only alternative. There is nothing behind the appearances, but this doesn't mean there is no more than the appearances. There is the dynamic depth in the appearance which is the appear*ance*. Because it is the appear*ance*, it is the thing itself (*not* the thing-in-itself) manifesting – we can say this about the appear*ance* but not the appearance. This is the dynamics of being which replaces the two-world theory that separates being from appearance.

The Illusion of Independent Existence

There can be no such thing as an entity that is absolutely independent, being what it is solely in terms of itself, without any relation to what is other than itself. Every distinction, in order to *be* a distinction, is necessarily a unitary act of {differencing/relating} – it is one 'movement' which goes in opposite directions simultaneously. Thus difference *without* relation is actually unthinkable, although we usually don't notice this and fall into the error of believing that we can think of distinction as just difference, because we begin at the end with *what* is distinguished instead of with the *act* of distinction itself. In this case we can appear to have a distinction which does not entail a relation because it is already 'too late'; what we are thinking of as a distinction is in fact the *separation* of what is already distinguished. A distinction which did not entail a relation would be an absolute distinction. Hegel points out that such an absolute distinction would be self-contradictory. Because it would not entail a relation we could not say what it distinguished. By annihilating the relation implied in the distinction it would annihilate the distinction itself. Thus an absolute distinction would not be a distinction at all.[23] Of course, we often do think of things as if they were separate and independent existences, and as an approximation it may often be admissible and useful to do so. The problem comes when we fail to remember that this is only an abstraction, and that in the concrete situation there are no such separate and independent existences.

The fundamental relations which any entity has to other entities are sometimes said to be internal to that entity – i.e. other entities enter into the very constitution of 'what it is' – instead of being external to it as they would be if entities existed separately and independently. In other words, any entity is what it is only within a network of relations. So instead of being an atomic existence it is in fact holistic. When we think materialistically of the world as being 'made up' of separate and independent entities, which are like building blocks, then we really have got it backwards:

> The attempt to rationally reconstruct the world out of a collocation of 'bits' contingently related to one another is as futile as the attempt to appreciate a symphony by sounding each note in isolation and then imagining a relation among them.[24]

These separate 'building blocks' only seem to be such when we begin 'downstream', whereas when we go 'upstream' we discover that the world is intrinsically holistic. So the question becomes, not how do entities which are separate and independent become related to one another, but how does it seem that there are such separate and independent entities in the first place? We find the answer when we go 'upstream' into the primary act of distinction, where we discover that relation is intrinsic to distinction, and that things only appear to be separate and independent when attention is focused 'downstream' on what is distinguished.

2. Goethe and Modern Science

I believe that a major obstacle standing in the way of our understanding of Goethe's alternative approach to the science of nature, is that we often have an inadequate understanding of the way that mainstream science developed historically. As a consequence, we have several misconceptions about science, and fail to realise that the direction taken by modern science is only one possibility. In the beginning there are always more possibilities than the one actually taken. The choice which is made opens the door into the way that is followed, but at the same time it closes the door to other possibilities, which consequently withdraw into the background and are no longer noticed. They become invisible, but they do not cease to exist, and the time will come when unexpected consequences of the choice that was made, will begin to redirect attention to other possibilities that were not taken. This is in effect what Goethe did, and to understand this we must begin by becoming clearer about the pathway taken by mainstream science. What we find is that there is a strong tendency towards underestimating the formative influence of the mathematical style of thinking in the development of modern science, whilst at the same time overestimating the influence of empiricism.

The Beginnings of Modern Science

That Galileo's aim was to reject and replace the Aristotelian approach to science is now part of what 'everybody knows'. But, true as this certainly is in several respects, it is not completely true. If we look at the development of seventeenth century science in a more comprehensive historical context, instead of just treating it as a historically local event which swept away all that had gone before, we will find a remarkable continuity in the ideal of scientific methodology, going back through Robert Grosseteste and Roger Bacon in the thirteenth century to earlier medieval thinkers and Arab scientists and commentators, and ultimately back to Aristotle's *Posterior Analytics* as the founding text on which science is based. Rather than sweeping all this away to begin anew, Galileo's account of his work on the science of motion fits in with it in an exemplary fashion, albeit in his own individual way. This is clearly important for gaining a fuller understanding of the meaning of science.[1]

Before the twelfth century science was entirely empirical, and this was its limitation. Although based on observation, it could not yet get beyond the rule-of-thumb methods of the practical crafts to the stage of becoming a 'theoretical science offering rational explanations of the facts of experience'.[2] This is sometimes referred to today as the difference between 'cookbook science' and 'explanatory science'.[3] The crucial step from purely empirical science to what can properly be called rational-empirical science – which is modern science – became possible during the twelfth century when logicians were able to make use of Aristotle's *Posterior Analytics*:

> What these logicians did was to recognise the distinction between experiential knowledge of a fact and rational or theoretical knowledge of the cause of the fact, by which they meant knowledge of some prior principles from which they could deduce and so explain the fact.[4]

This is the key distinction which derives ultimately from Aristotle:

> According to Aristotle, scientific explanation was a twofold process, the first being inductive and the second deductive. The investigator must begin with what was prior in the order of knowing, that is, with facts observed through the senses, and he must ascend by induction to generalizations or universal forms or causes which were most remote from sensory experience, yet causing that experience and therefore prior in the order of nature. The second process in science was to descend again by deduction from these universal forms to the observed facts, which were thus explained by being demonstrated from prior and more general principles which were their cause.[5]

Although this double movement, from experience to theory and from theory to experience, is formulated by Aristotle expressly for science (*Posterior Analytics*), the root of it goes back to Plato (who was not concerned with science). It is there in the Divided Line in the *Republic* (509d–511e). Plato was very much at the centre of the remarkable advance in mathematics which was taking place in Athens, and his philosophy was clearly strongly influenced by this in several ways. But among all the mathematical discoveries that were made about geometry and number, the most important one from the perspective of later developments was the *methodological* discovery of the possibility of *deductive proof*. It is this which is the hallmark of classical Greek mathematics, and which enabled Plato to insist that mathematics is not some kind of empirical investigation – a distinction that was later to have such an influence in the direction taken by western thinking. Aristotle's deductive logic, which he describes in *Prior Analytics*, was derived from the kind of reasoning which he observed being practised by the mathematicians – so that deductive logic has been called 'the child of mathematics'.[6] This great methodological discovery of the Greeks flowered two generations after Aristotle in Euclid's *Elements of Geometry*, a work which has had an almost inestimable influence on western thinking since it was introduced from the Arabs in the Middle Ages. This work gives the model for the deductive movement in science from theory to experience, whereby the phenomenon that has been investigated can then be deduced from a first principle that has been discovered by induction.

What was additionally introduced into the 'double way' in the Middle Ages was the use of experiment, both as an aid to discovery and as a means of verification/falsification. With this the methodological revolution of modern science began, in Oxford with Grosseteste and others who followed him, who included Roger Bacon, and in Paris with Albertus Magnus and others. Crombie concludes:

> The conception of the logical structure of experimental science held by such prominent leaders as Galileo, Francis Bacon, Descartes, and Newton was precisely that created in the thirteenth and fourteenth centuries.[7]
> The history of the theory of experimental science from Grossesteste to Newton is in fact a set of variations on Aristotle's theme, that the purpose of scientific inquiry was

to discover the true premises for demonstrated knowledge of observations, bringing in the new instrument of experiment and transposing into the key of mathematics.[8]

This is precisely what we see Galileo doing with the science of kinematics on the Third Day in *Dialogues Concerning Two New Sciences*. He describes the two parts of the double procedure in terms which had been in use since the thirteenth century. Einstein came to realise that the standard interpretation we have all been taught misrepresents Galileo. When invited to contribute a Foreword to Stillman Drake's translation of Galileo's *Dialogue Concerning the Two Chief World Systems*, he wrote:

> It has often been maintained that Galileo became the father of modern science by replacing the speculative, deductive method with the empirical, experimental method. I believe, however, that this interpretation would not stand close scrutiny ... To put into sharp contrast the empirical and the deductive attitude is misleading, and was entirely foreign to Galileo ... The antithesis Empiricism vs. Rationalism does not appear as a controversial point in Galileo's work.[9]

Nor does it appear in the work of Francis Bacon, Descartes, or Newton. In each of these, as well as Galileo, we find methodologically the same double procedure that had been developed since Grosseteste at the end of the thirteenth century, and which goes back ultimately to Aristotle. The confusion which has arisen about this is because of the failure to distinguish between the methodology and the results of science. Galileo and the others found plenty to disagree with in the dogmatic assertions of the Aristotelians, but their rejection of specific results is not the same as rejecting the overall methodology. However, these have been confused, with the consequence that the deeper continuity of science from Aristotle onwards has been eclipsed and a false image of the development of modern science has been established.

This idea of theory-based scientific explanation was the major innovation which transformed science from being *only* empirical into the *rational* empirical form which we recognise today as being characteristically 'scientific'. This has been extraordinarily successful, but it does have the effect of shifting attention away from the phenomenon, with the result that the phenomenon itself begins to take second place in favour of the theory. Paradoxically, science becomes theory-centred instead of phenomenon-centred. This is particularly the case when mathematics begins to play a fundamental role in science. It is evident that, by its very nature, mathematics takes us away from the concrete into abstraction. But this in itself does not necessarily undermine the value of the sensory. We can count, measure, weigh, and so on, without this implying that those aspects of nature which can be quantified are in any way more real, or more fundamental, than those qualities which cannot be readily quantified in the same way – such as colour, for example.

We can discover mathematical proportions and relationships in nature which lead us away from the diversity of sensory appearances towards the discovery of a unity which is more abstract inasmuch as it does not depend on the differences between specific instances, but is the very same in all cases. This led to the remarkable idea that there are universal laws of nature, which is 'rather surprising, for nothing is less evident in the variety of nature than the existence of universal laws'.[10] But although the mathematical style of thinking in physics leads us away from the experience of the senses as such, there is no intrinsic reason why this should make us think of the world as experienced through the senses as being *inferior* in any way to the relationships in nature discovered by means of mathematics. There is no reason why we should think of these mathematical relations as being more than a facet or aspect of the appearance of nature (it is more dynamic than a perspective – it is nature manifesting mathematically), or any reason why we should take the mathematical as being in some way more fundamental than the sensory aspect of nature. As Aristotle recognised, reality can be considered under various aspects without having to consider that one is more fundamental than another. This is emphasised by Heidegger in the plural realism of his existential philosophy of science, according to which in the words of Hubert Dreyfus:

> Reality can be revealed in many ways and none is
> metaphysically basic ... And just because we can get things
> right from many perspectives, no single perspective is the
> right one.[11]

If we think this looks like relativism, it is because we have presupposed already (even though we may not be aware of it) that only one way of describing nature can correspond to 'the way things really are':

> But for Heidegger ... Since no one way of revealing is
> exclusively true, accepting one does not commit us to
> rejecting the others.[12]

But unfortunately, this is not the kind of understanding that grew out of the way in which physics developed historically.

The Temple of the Sun

It is now widely recognised that the development of science is not determined solely by empirical and methodological factors, but that it is also a consequence of the influence of contextual factors – which include 'schools of thought'. This means that cultural-historical influences enter into the very form which scientific knowledge takes. In other words, science itself is intrinsically historical – which is referred to as the 'historicity' of science – and is not independent of cultural-historical influences which often determine what gets taken as fundamental – what is 'really real' – and therefore also what gets relegated to being only 'secondary'. Two very different schools of thought had such a formative influence on the development of modern science: the philosophy of Neoplatonism and the philosophy of Atomism. In their different ways both of these had the effect of relegating the sensory world to a secondary status, even promoting the idea that the senses are 'confused' and consequently not to be trusted. In one way or another, emphasis on the *inferiority* of the senses and the superiority of mathematics has been a dominant characteristic ever since, and against such a historical background it is not surprising that Goethe's re-investment of attention in the sensory phenomenon has not been appreciated.

We can see this exemplified clearly in the case of Copernicus. Why did he propose the heliocentric planetary system? It is often assumed that it must have been prompted by new observations of some kind – though it is difficult to see just what kind of empirical evidence could lead to such a major structural change of the cosmos as that initiated by Copernicus. In any case, there were no radically new observations, only improvements on earlier measurements. Another answer to the question which is often given simply assumes that the heliocentric theory gave a more precise account of the measured positions of the planets in the celestial sphere. But again this isn't true. To begin with at least, the theory of Copernicus was no better in this regard than the traditional geocentric theory. So why did Copernicus do it? We find the answer by reading what he said about it himself in the first part of his book, *De Revolutionibus Orbium Caelestium* (1543), as well as in the prefatory letter that he wrote to the Pope in an attempt to forestall the difficulties he anticipated. Putting the Sun at rest in the centre and moving the Earth, contrary to common sense experience, enabled Copernicus to achieve a far greater degree of *mathematical harmony* than was otherwise possible. He says:

> Thus assuming motions, which in my work I ascribe to
> the Earth, ... I have at last discovered that, if the motions
> of the rest of the planets be brought into relation with the
> circulation of the Earth ... The orders and magnitudes of all
> stars and spheres, nay the heavens themselves, become so
> bound together that nothing in any part thereof could be
> moved from its place without producing confusion of all the
> other parts and of the Universe as a whole.[13]

There is 'a clear bond of harmony in the motion and magnitude of the Spheres such as can be discovered in no other wise'.[14] In other words, the achievement of mathematical harmony by putting the Sun in the centre and moving the Earth, means that the 'Solar System' (as it now becomes) is a *holistic* system, instead of the assemblage of disparate parts that Copernicus claims is all that it was before his innovation.

But the Sun was not placed in such a prominent position only because of the mathematical advantage this brought. It was an expression of the importance which the Sun has in the Neoplatonic philosophical tradition. This philosophy – which as the name suggests derives ultimately from interpretations of the writings of Plato – began to have a cultural impact again at the time of the Renaissance. There are different currents in this school of thought, but in the Renaissance what was of interest to Copernicus and others was the emphasis on the importance of finding *simple* geometrical and arithmetical proportions in nature and the cosmos, and the role of the Sun as the creative source of light and life in the Universe. This is why we find Copernicus writing eulogistically:

> In the middle of all sits Sun enthroned. In this most beautiful
> temple could we place this luminary in any better position
> from which he can illuminate the world at once? He is rightly

called the Lamp, the Mind, the Ruler of the Universe ... So the Sun sits as upon a royal throne ruling his children the planets which circle round him.[15]

When he refers to the Sun as being in the middle of 'this most beautiful temple', we should not take this as just hyperbole. What Copernicus says about the heliocentric system of planets being such that 'nothing in any part thereof could be moved from its place without producing confusion of all the other parts and of the Universe as a whole', is strikingly similar to what the Renaissance architect, Palladio, later said about the beauty of a temple: that it will result 'from the correspondence of the whole to the parts, of the parts among themselves, and of these again to the whole; so that the structure may appear as an entire and complete body, wherein each member agrees with the other and all members are necessary for the accomplishment of the building'.[16] Copernicus meant what he said to be taken seriously: when the planetary system is seen from the Sun instead of the Earth, mathematical proportions are discovered which give the heliocentric system the aesthetic form of a temple. It is an important feature of the Neoplatonic approach that these proportions are not visible in the cosmos as it is experienced by means of the senses. What the mathematical Neoplatonist discovers, obscured by the senses and therefore 'hidden behind the appearances', is that the system of the central Sun and planets has the mathematical form of a heavenly temple for the presence of the living God.

The Neoplatonic influence is shown very clearly in Kepler's determination to establish the Sun at the centre of the planetary system. He searched the data on planetary positions to discover the simple geometrical and arithmetical relationships in the movements of the planets which we now call Kepler's laws. But this was no straightforward empirical procedure – as if he could just find the mathematical proportions there in the data – because he was guided throughout by the idea that the Sun *must* be in the centre, this being the position where the 'Most High God' would choose to dwell if 'he should be pleased with a material domicile'.[17] It is because he insisted on interpreting the data *in the light of this idea,* that he was led in the end to introduce irregularity into both the shape of the orbit (not quite circular but slightly elliptical) and the movement of the planet (not quite constant speed). But his determination to find simple mathematical proportions in the planetary motions was sustained by his Neoplatonic belief that he was searching for relationships which are *transcendent* to nature – i.e. beyond and ontologically superior to nature as it appears to the senses. In the Christianised Neoplatonism of the time, it was only a short step to identifying these mathematical laws of nature (as they were called subsequently) with the 'thoughts of God'. This is what Kepler says about geometry:

> Why waste words? Geometry existed before the Creation, is co-eternal with the mind of God, *is God himself* (what exists in God that is not God himself?); geometry provided God with a model for the Creation and was implanted into man together with God's own likeness – and not merely conveyed to his mind through the eyes.[18]

This belief that mathematics leads to the discovery of relationships which transcend the appearances – i.e. that they are *beyond* nature as we encounter it through the senses, and not simply within it but not accessible to the senses as such – is usually traced back to the philosophy of Plato. This may be unwarranted – or at least not without a great deal of qualification – but 'Platonism' is the name that is usually given to the two-world theory which *separates* (and not simply distinguishes) the intelligible (which in this case is the mathematical) from the sensible.[19] In this metaphysical picture the two worlds are each given a different ontological status, with the intelligible being superior or 'higher' and the sensible inferior or 'lower'. In the present context, this means that the mathematical ratios and proportions are conceived as being in a higher ontological realm which is *separate* from the lower ontological realm of the phenomena as they appear to the senses – an immutable realm of transcendent mathematical forms *separate* from the realm of changing sensory appearances. It is this transcendent mathematical world behind or beyond the world of sensory phenomena – and which is the real being of the phenomena (even though it is separate from them!) – that it is supposed is discovered by the science of mathematical physics. As one modern philosopher, Gary Madison, has recognised: 'Metaphysics finds its ultimate expression in modern, mathematical physics' and hence 'Metaphysics is alive and well and lives on in modern physics'.[20] This is the last place we would expect to find it.

Galileo and the New Science

With this 'Platonic' background to the development of modern physical science, the senses were relegated to a secondary ontological status and this gave rise to the unwarranted view that the senses are inferior. Certainly the world of the senses does not explain itself, and consequently the senses are not sufficient in themselves for a scientific explanation of phenomena. But this does not imply the *inferiority* of the senses, although we can easily see that the temptation to think this way is almost an inevitable consequence of the two-world theory. But it went further than this, when the supposed inferiority of the senses came to be seen as implying that the senses are untrustworthy, that they deceive us, leading us into error and illusion. We can see how easily this seems to be so in the case of the Copernican transformation. The Earth is moving but our senses tell us that it is at rest; we see the Sun, Moon, stars and planets, moving across the heavens daily, but it is the Earth turning; we see retrograde 'loops' in the motion of the planets as seen against the background of the stars, but this is an 'illusion' resulting from the movement of the Earth compounded with the motion of the planets. The truth is only discovered by mathematical thinking, which reveals to us the ratios and proportions that are 'hidden behind the senses' and which alone explain the world as it appears. Galileo expresses this attitude very clearly:

> I cannot sufficiently admire the eminence of those men's wits, that have received and held it to be true, and with the sprightliness of their judgements offered such violence to their own senses, as that they have been able to prefer that which their reason dictated to them, to that which sensible experiments represented most manifestly to the contrary.[21]

Galileo recognised that an entirely new science of motion would be needed for the movement of bodies on the Earth if the Earth itself were moving. He said, through the mouthpiece of Simplicius in *Dialogue Concerning the Two Chief World Systems*: 'The crucial thing is being able to move the Earth without causing a thousand inconveniences'. For example, since the Earth is travelling from west to east at great speed, we would expect a ball dropped from a tower to reach the bottom well to the west of the tower. In fact it drops straight down to land at the foot of the tower – just as it would be expected to do if the Earth were not moving. So here is one inconvenience that doesn't happen. The natural thing to do would be to conclude from this that the Earth *doesn't* move. But if we are going to insist that, contrary to the evidence, the Earth *does* move, then we are going to have to explain why this inconvenience doesn't happen. Goethe wrote: 'The greatest art in theoretical and practical life consists in changing the *problem* to a *postulate*; that way one succeeds'.[22] He particularly admired Galileo for doing just this. Galileo solves the problem of how the Earth can move without causing a thousand inconveniences, by *postulating* that the Earth moves without the inconveniences happening! He has to change the whole of the physics of motion to accommodate to this postulate. This requires a fundamental change in the very idea of motion itself so that Galileo can conceive the idea of *inertial* motion.[23] His answer is that this inconvenience (or indeed any of the others) doesn't happen because 'keeping up with the Earth is the primordial and eternal motion ineradicably and inseparably participated in by this ball as a terrestrial object, which it has by its nature and will possess forever'.[24] In other words, in keeping up with the movement of the Earth it is just 'doing what comes naturally'. So as the ball is falling it will continue to move with the Earth, which means that it will reach the ground at the bottom of the tower – just as it would have done if the Earth itself had not been moving. Thus by introducing a new conception of motion – the idea of inertial motion –Galileo can accommodate the fact that a body seems to move on the moving Earth in exactly the way that it would if the Earth itself didn't move. But in doing this he goes against the experience of the senses and the 'common sense' based thereon:

> Not the least of what sensible experience showed men –
> or perhaps seemed to show them before Galileo instructed
> them to interpret experience otherwise – was that force is
> necessary to keep a body in motion. Indeed, where is the
> experience of inertial motion? It is nowhere.[25]

But this is not the only step which Galileo took that leads away from the sensory into the mathematical. He was also instrumental in introducing the ancient Greek philosophy of Atomism into physics.[26] In doing so he made a fundamental division between those qualities of nature which can be quantified directly and those which cannot, such as colour. It is evident that there is a *methodological distinction* between those qualities of nature that can be mathematised and those which cannot. But Galileo went far beyond this to introduce an *ontological division* – although in doing so he was only going in the direction already taken by the Greek atomists (although without the mathematical imperative in their case). This is what Galileo himself says in *Il Saggiatore (The Assayer)*:

> Now I say that whenever I conceive any material or corporeal substance, I immediately feel the need to think of it as bounded, as having this or that shape; as being large or small in relation to other things, and in some specific place at any given time, as being in motion or at rest; as touching or not touching some other body; and as being one in number, or few, or many. From these conditions I cannot separate such a substance by any stretch of my imagination. But that it must be white or red, bitter or sweet, noisy or silent, and of sweet or foul odour, my mind does not feel compelled to bring in as necessary accompaniments. Without the senses as our guides, reason or imagination unaided would probably never arrive at qualities like these. Hence I think that tastes, odours, colours and so on are no more than mere names so far as the object in which we place them is concerned, and that they reside only in the consciousness. Hence if the living creature were removed, all these qualities would be wiped away and annihilated. But since we have imposed upon them special names, distinct from those of the other and real qualities mentioned previously, we wish to believe that they really exist as actually different from those.[27]

Galileo takes those qualities which cannot be directly mathematised out of nature altogether and relocates them entirely within the human being. There is now, not just a distinction, but a division between what later came to be called (by the philosopher John Locke) 'primary' and 'secondary' qualities. The primary qualities are those which are quantitative, and which alone are considered to be real in nature. The secondary qualities are those which, although appearing to be real, are nothing more than the effects of the primary qualities on the senses. The result is the subjectivisation of the secondary qualities with the consequent depletion of nature. Needless to say, there is a good deal of *philosophical* confusion here, which for the most part we still live with today, and which doesn't even begin to get cleared up until we come to the phenomenology of the lived body and the life-world in the twentieth century. The upshot of what Galileo did is that the 'illusion of the senses' is now compounded. Not only is reality different from the sensory appearances because it is mathematical, but also the sensory appearances themselves seem to be even more untrustworthy, because much of what they are telling us about the world turns out not to be in the world at all but only in our subjective experience. So it seems that the senses deceive us even more than we imagined. We can see a bifurcation beginning to emerge here, according to which the real world is outside of humanity, so that *ipso facto* humanity is now outside of the real world. As Burtt observes: 'the stage is fully set for the Cartesian dualism – on the one side the primary, mathematical realm; on the other the realm of man'.[28]

Descartes Seeks Foundations

René Descartes took this anti-sensory mathematical ontology to an extreme. He was greatly impressed by the work of Galileo, but believed that 'he has built without a foundation'.[29] His aim was to provide a metaphysical underpinning for the new science of mathematical physics which would distinguish as clearly and completely as possible between the sense-based conception of nature and the mathematical conception. This would make it evident that, although the latter is far removed from the experience of the senses, there are compelling grounds for believing that it is the mathematical approach which gives us the true knowledge of nature, whereas what the senses show us is an illusion. He went even further than others in this direction – although it could be said that he was only drawing to a conclusion what they had begun:

> The cornerstone of the entire edifice of his philosophy of nature was the assertion that physical reality is not in any way similar to the appearances of sensation. As Copernicus had rejected the view of an immovable earth, and Galileo the common sense view of motion, so Descartes now generalized the reinterpretation of daily experience.[30]

The way that Descartes tried to do this – which at the time also had a theological purpose – was to show that 'the human mind was constituted by God to enjoy perfect certainty about material things when conceiving them mathematically'.[31] The upshot was, he believed, that when the human mind is occupied with mathematical physics it is doing the very thing for which it was created by God. It would be difficult to imagine a better warrant for mathematical physics than this! No matter how strange it may seem to us now, 'we must remember that the whole course of modern science has been run, not be returning to the earlier philosophy of nature, but by following the path he chose'.[32]

It was in the course of trying to give grounds for believing that it is the mathematical approach that gives us true knowledge of the world, and not the senses, that Descartes introduced the ontological dualism for which he is famous. He tells us that he is going to 'apply myself seriously and freely to the general destruction of all my former opinions' (First Meditation). He feels this is the necessary step which has to be taken first if he is to 'begin afresh from the foundations'. Indeed, Descartes is widely believed to have done just this. It is therefore surprising to see how the entire edifice of his thought is based from the outset on one of the key concepts of the medieval scholastic philosophy which he sought to replace. This is the concept of 'substance', which comes originally from Aristotle. Because modern readers are not familiar with this notion it easily gets overlooked, and the modern meaning of 'substance' is substituted instead. But this is not the same thing. In medieval thought, something is a 'substance' if and only if it can exist entirely independently of anything other than itself. Each substance has a distinct essence that is uniquely its own. For Descartes there are two such substances. These are the world and human being, each of which has its own characteristic essence. The essence of the world or matter, which includes the body, is 'extension', whereas the essence of human being is 'thinking'. He calls these *res extensa* and *res cogitans*, and *by definition* they are entirely independent of one another. Here we have the Cartesian dualism according to which reality divides exclusively into mind and nature (which is thereby now reduced to matter), which are entirely distinct and separate from each other. Furthermore, since the body is extended it belongs with *res extensa* and is therefore part of the world – and hence is excluded from human being whose essence is thinking. So the Cartesian bifurcation of reality is at the same time the dualism of mind and body (Descartes himself never used the term 'consciousness', which was introduced later by Locke).

To bring out more clearly the astonishing lack of relationship between human being and the world which is implied by Descartes' two-substance ontology, we should focus, not so much on the epistemological problem of how we can know the world from which we are cut off (which has been the main focus of subsequent western philosophy on account of the emphasis on science), but more on the ontological consequence of proposing that the thinking mind is self-sufficient and self-contained. This consequence is described very clearly by David Cooper:

> Descartes tries to show that my experiences could be just as they are even though there exists no world for them to be experiences of. And what is disturbing in this is not the worry that perhaps there really is no world, but the sense that I am a self-enclosed realm, 'cut off' in logical isolation from the world. If I could exist despite the absence of things and other people, then it cannot be *essential* to my being that I have a body and am in the company of others.[33]

The puzzle as to why Descartes does this – especially after he has told us that he intends to sweep away all previous concepts and begin again – is resolved when we remember the wider context of his enterprise. His aim is to give a foundation to mathematical physics which is consistent with the doctrine of the Church. Look at the full title of the *Meditations,* with its reference to 'the real distinction between the soul and the body of man'. He is going to show that the aptitude for doing mathematical physics is an aptitude of the soul which does not depend in any essential way on the body. There is no need here to go into the details of how Descartes believes he can achieve this. The essential point is that he introduces a dualism based on the traditional concept of substance which is entirely congruent with the teaching of the Church concerning the immortal soul and mortal body. In this way he hopes to show that the new mathematical philosophy of nature is in harmony with mainstream Christianity, and hence that it could replace the Aristotelian philosophy in the synthesis with Christianity that Aquinas had produced, and which had become the official philosophy of the Church. Taking this background into account, we can see that Descartes effectively assumes at the outset what he purports to conclude by the exercise of reason:

> Against this background it is possible to make sense
> of Descartes's otherwise arbitrary distinction between
> purely intellectual capacities on the one hand, and
> body-presupposing capacities of sense experience and
> imagination on the other. This can then be seen as a division
> of capabilities into those we can share with God, and in
> virtue of which we can have something like his objective
> understanding of reality, and those we do not share with God
> and that are not necessary for objective understanding.[34]

This is really where Descartes grounds certainty. It seems to him that only a mind that is completely distinct from the body could achieve knowledge which does not depend on the experience of the senses. The guarantee that mathematical physics gives us the truth is that when we are doing it we are using only the capacities of the soul, which we share with God, and not depending on the bodily senses. So who could doubt that the mathematical way does lead us to the truth beyond the illusion of the senses?

Descartes takes this very seriously, so much so that it enters into the very form which scientific knowledge takes for him. One of the first tasks that a corpuscularian physicist must undertake is to find the mathematical form of the laws of impact when one corpuscle collides with another. Descartes considers this in detail and proposes seven laws of impact, almost all of which seem to be false – the consensus being that the correct laws were discovered later by Huygens. Sometimes what Descartes says seems so evidently wrong that it is difficult to understand why he proposed it in the first place. For example, his fourth law of impact states: If body A is at rest and is larger, however slightly, than body B, then no matter with what velocity B strikes A, A will never be set in motion and B will be reflected back in the direction from which it came. Hübner, who has investigated Descartes' laws of impact in some detail, comments that everyone will reject this 'since it contradicts even the slightest experience', and yet 'Descartes himself is not bothered at all by this, though he must have recognised it'.[35] The resolution of this puzzling state of affairs comes with the realisation that Descartes is *not* thinking in terms of the concepts of motion that later emerged in Newton's *Principia* – and which we assume apply in Descartes' case, but in fact don't. What Hübner discovers is that:

> The force that Descartes holds to be operative in the
> interaction of impinging bodies has nothing to do with
> momentum as we understand it. It relates neither to inertial
> masses nor to a velocity dependent upon a human time
> measurement and possible perceptions relating to a body that
> is only moved in some relative sense. Rather here we see that
> *Descartes' laws of impact describe fundamental occurrences of*
> *nature as if seen from the standpoint of God,* that is, occurrences
> related to a duration and motion *in rebus* or *sub specie*
> *aeternitatis.* Thus these laws are part of a 'Divine Mechanics'.[36]

The nineteenth century term 'celestial mechanics' – used to describe the attempt to get better approximations to the planetary orbits by perturbation theory – literally takes on a theological meaning in the case of Descartes! What this shows is that Descartes really did believe that the intellectual capacity we share with God (in his view) can take us beyond the prejudices of the senses (as he saw them) to discover by the light of reason the way that the mechanics of the universe is seen by God its Creator:

> The invisible world, underlying the visible and alone serving
> as a ground for the interpretation of the latter, is known
> by an indubitable reason that sees through the sensible to
> its true cause and knows itself to be one with the light of
> divine revelation. And it is thus precisely for this reason that
> Descartes evinces his provocative disinterest in what is clearly
> perceived by the senses and indeed even challenges this, as is
> particularly evident in the fourth rule of impact.[37]

This is therefore the absolute truth because it is how things are for God, which is the aim and culmination of Descartes' whole enterprise – and which we can now see makes sense as a whole.

However, it turns out that this metaphysical-theological under-pinning of mathematical physics is not the only reason why Descartes introduced such a radical divorce of mind from nature. There is a further factor in the cultural-historical context at the time, one which historians have only recognised more recently and which also needs to be taken into account. As well as opposing the Aristotelian philosophy of the Establishment, the new mathematical-mechanical philosophy also had to compete with another prevalent philosophy of nature in the fifteenth and sixteenth centuries. This is often called Renaissance Naturalism (it is also sometimes referred to as the Hermetic Tradition). This philosophy of nature also claimed to replace the medieval synthesis of Aristotelianism with Christianity, but by a synthesis of Christianity with the hermetic philosophy, as developed by Ficino, Pico della Mirandola, and others in the Neoplatonic Tradition, supplemented by the discovery of the Corpus Hermeticum and the influence of the Cabala.[38] This may seem very strange to us today, because we are the children of the outcome in favour of the mathematico-mechanical philosophy, which has done so much to shape the modern world and our attitude towards nature.

Descartes and others at the time, notably Mersenne and Gassendi, saw Renaissance Naturalism as a philosophy of nature that they had to repudiate once and for all. We can see how Descartes contributed to this when we realise that, for Renaissance Naturalism, nature was something living and seeing. As well as being physical, nature also had psychic qualities. So as well as being material, there was something mind-like in nature which was its active principle. The natural philosopher's task was to understand nature by drawing it into himself, internalising the psychic in nature in his own psyche, so that what was mind-like in nature could come into being in his own mind. The active principle which was the living inwardness of nature (its psyche), could therefore manifest itself directly in the inwardness of the philosopher (his psyche), and consequently he would literally have a deeper understanding of nature than could be achieved by external means alone. So by dividing reality into two substances (in the traditional sense of this term) which are by definition mutually exclusive – *res extensa* and *res cogitans* – Descartes is thereby excluding any mind-like or psychic qualities from nature, and reducing it to a purely material nature that consists of nothing more than the mechanical collisions of inert material particles. Westfall points out that:

> Descartes' choice of the passive participle, *extensa,* in contrast to the active participle, *cogitans,* which he used to characterize the realm of spirit, served to emphasize that physical nature is inert and devoid of sources of activity of its own.[39]

If the *only* active principle is the human mind, there now being nothing psychic in nature, then the ground is cut from under the feet of the Renaissance nature philosopher. He cannot internalise what is mind-like in nature in his own mind because there is no such thing. Instead, nature is now conceived as being totally devoid of mind-like qualities. Since such qualities are now conceived as being restricted exclusively to the human mind, it follows that, far from internalising nature, the mathematico-mechanical philosopher *externalises* it by separating it completely from the mind. Now we have the condition of objectivity which is the necessary condition for the very possibility of modern science. This is what Descartes achieved. However, we can begin to get the sense that, immensely successful though it has been, this has been achieved only by the drastic reduction of nature – and along with it an equally drastic reduction of the human body – to something which is effectively dead.[40]

Yet if we stop to ask ourselves what our experience would be like if we were such dualistic beings, we soon realise that it would be an utter nightmare. We would find ourselves permanently in the state described by Oliver Sacks in the case of 'The Disembodied Lady'.[41] At the age of twenty-seven, Christina lost all proprioception – the sense we have of ourselves from the movable parts of our body (muscles, tendons, joints) by means of which 'we feel our bodies as proper to us, as our "property", as our own'. She found she could not stand – unless she looked down at her feet. She could not hold anything in her hands – which wandered about unless she kept her eye on them. Although her vocal posture had gone, she managed to say, in a ghostly flat voice: 'Something awful's happened. I can't feel my body. I feel weird – disembodied'. When she was told how the sense of the body is given by proprioception, she said:

> This 'proprioception' is like the eyes of the body, the way the body sees itself. And if it goes, as it's gone with me, *it's like the body's blind*. My body can't 'see' itself if it's lost its eyes, right? So *I* have to watch it – be its eyes. Right?

So this is what she had to learn to do. She had to monitor herself by vision, looking at each part of her body with careful attention as it moved. Gradually her movements became less clumsy and artificial and began to seem more 'natural', though still wholly dependent on the use of her eyes. With time 'the normal, unconscious feedback of proprioception was being replaced by an equally unconscious feedback by vision, by visual automatism and reflexes increasingly integrated and fluent'. But this substitution did not result in her movements ever becoming entirely natural again – there was always something artificial and even overcompensated about them – and it had no effect whatsoever on her sense of being disembodied: 'she continues to feel, with the continuing loss of proprioception, that her body is dead, not-real, not-hers – she cannot appropriate it to herself'. Her body has no sense of itself, but she has developed ways of imitating normal life. Sacks comments that 'in an extraordinary way, she has both succeeded and failed. She has succeeded in operating, but not in being'. Christina herself puts it more dramatically:

> It's like something's been scooped right out of me, right at the centre ....that's what they do with frogs, isn't it? They scoop out the centre, the spinal cord, they *pith* them ... That's what I am, *pithed*, like a frog ... Step up, come and see Chris, the first pithed human being. She's no proprioception, no sense of herself – disembodied Chris, the pithed girl!

This breakdown of the normal situation gives us a picture of how we would all be if Descartes' dualistic ontology really did depict what it is like to be a human being. In this dualism the body is depicted as being entirely external to the mind, which is where 'I am, I exist', so that it would be experienced as something alien to me, but which always accompanies me as an object to which I am mysteriously attached. But in terms of this mind-body dualism, it would not be possible for us to *experience* this body, and so we would all be pithed human beings.

Obviously Descartes was well aware that we do not in fact experience ourselves like this, as a mind/soul joined to a mechanistic object. He conceded that our relationship with our body is not like that of a ship's pilot who observes the condition of his ship. But he was at a loss to know how to account for it. Convinced that in order to give a metaphysical underpinning for mathematical physics – as well as for the concomitant theological and other reasons we have seen – ontological primacy must be given to the mind/soul over the body, he nevertheless recognised that this was at odds with ordinary life experience. The best he could do to explain the illusion (as it must be for him) of our ordinary experience was to point out that it is beneficial

47

for our survival. It is better for us to feel that we are intimately fused with our bodies, instead of being minds joined to bodies that are really distinct from ourselves, because this illusion encourages us to avoid harm. For example, if I put my hand in the fire I may easily just leave it there if it is my experience that the pain is in a body which is separate from me. In this case my experience would be that the pain is outside of me in the world – this being where the body that accompanies me belongs. It is to avoid such injurious consequences of ontological dualism that Descartes believes it is necessary to have the illusion that the pain is *mine,* and not in a body which is external to me. He suggests that this illusion is an example of God's benevolence, because whereas God would not deceive us, in this case his benevolence allows us to live the illusion of ordinary life (the pain is mine) for the sake of our own protection. This is the extent to which Descartes has to go to explain our ordinary life experience, so that he can give what he sees as convincing grounds for believing in the truth of mathematical physics which is so contrary to the experience of the senses.

It is astonishing how such a back to front philosophy came to dominate the modern western mind. Although the difficulties it engenders soon became apparent, it nevertheless prevailed. This is especially the case in science, which historically was the matrix in which this philosophy emerged, and we can still see it today deeply entrenched in the traditional epistemology of cognitive science – in spite of claims to the contrary. Yet from the beginning of the twentieth century every aspect of this philosophy has been repudiated by phenomenology. For the traditional Cartesian position, consciousness is a subjective 'container' closed off from the world, in which ideas appear that function as representations of whatever is 'out there' in the objective world. This representational theory of perception, as it is called, has bedevilled philosophy and the sciences alike with its radical separation of consciousness from the world, and yet right from its beginning phenomenology shows us that the notion that consciousness is closed in on itself is completely at odds with lived experience. We discover instead the *intrinsic openness* of consciousness towards the world. This directionality – which is called 'intentionality' – is 'not an external relation that is brought about when consciousness is influenced by an object, but is, on the contrary, an intrinsic feature of consciousness.'[42] This intentionality which is the openness of consciousness towards the world, is often expressed by saying that consciousness is always 'conscious of' – which means that there cannot be consciousness without what it is 'conscious of'. So in this sense, we can say that, far from being self-enclosed, the very nature of consciousness is such that the world is already included within it.

One of the most remarkable and unexpected outcomes of phenomenology has been the rehabilitation of the body. The shift of attention away from the body as an object to the body as lived, is first described by Husserl in the second volume of *Ideas*, and subsequently developed further by Merleau-Ponty in *Phenomenology of Perception*.[43] It is through recognising the primacy of the incarnate subjectivity that is the living body – as opposed to the body as an object – that Merleau-Ponty's phenomenology can show us the way beyond the Cartesian mind-body dualism. This in turn opens the door to the rediscovery of living nature – not just organic nature, but the livingness of nature itself. The lived body is the sensing body and as such it is the organ through which we can encounter the sentience of nature. We can experience nature as living presence, instead of an object standing over against us. The very notion of the sentience of nature wouldn't make sense in physics (which does *not* mean that physics is wrong). But just as, according to Descartes, mathematical physics takes us 'out of' the body and separates us from nature, so the lived body can bring us into the presencing of nature. Such an encounter would be an impossibility within the framework of modern science, and yet it is only by awakening to this that we will really understand what is at stake in our relationship to the natural environment, and at the same time begin to wake up from our enthralment by the artificial world of technology.[44]

Newton and the Mathematical Physics of Nature

The mathematical philosophy of nature was brought to fruition by Isaac Newton in his extraordinary masterwork, *The Mathematical Principles of Natural Philosophy* (1687). Here we find for the first time what we recognise today as mathematical physics. Although he was nominally a mechanical philosopher, in the *Principia* (as it is usually called) he goes well beyond the mechanical philosophy. As well as atoms in motion, interacting mechanically by colliding with one another, he introduced the notion of forces which act between bodies that are not in contact. He introduced the idea of such 'forces acting at a distance' in an attempt to understand chemical interactions, the cohesion of bodies, and the capillary effect, but the most well-known instance is the universal force of gravitational attraction. This notion of a force acting at a distance clearly does not fit the mechanical philosophy. Newton himself believed that, far from being contrary to the mechanical philosophy, such forces should be seen as completing it. So the mechanical philosophy should be based, not just on the 'two catholic principles' (as Boyle called them) of matter and motion, but on three principles: matter, motion, and force – where the latter is not just the mechanical kind of force, as in a collision, but what in terms of the mechanical philosophy is strictly speaking not a mechanical force at all. Newton maintained that he had extended the mechanical philosophy to make it more comprehensive, but others were not convinced by this audacious step, and accused him of returning to the Renaissance Naturalism which the mechanical philosophy was designed to overcome. Newton responded by trying to show how a more orthodox mechanical explanation could be given for forces acting between particles that are not in contact – as in magnetism and gravity – but eventually he abandoned the attempt.

From the eighteenth century onwards, gravity began to be thought of as a 'property of matter', as if it were an attractive force inherent to matter. This is how it is taught in physics in schools today. But this is not what Newton thought. Koyré says, somewhat optimistically, that 'it is – or should be – a well-known fact that Newton did not believe in attraction as a real, physical, force'.[45] In his letter to Richard Bentley (written five years after the publication of the *Principia*), Newton says:

> You sometimes speak of gravity as essential and inherent to matter. Pray do not ascribe that notion to me, for the cause of gravity is what I do not pretend to know and therefore would take more time to consider of it.
>
> It is inconceivable that inanimate brute matter should, without mediation of something else which is not material, operate upon and effect other matter without mutual contact, as it must be if gravitation, in the sense of Epicurus, be essential and inherent in it. And this is one reason why I desired you would not ascribe innate gravity to me. That gravity should be innate, inherent and essential to matter, so that one body may act upon another at a distance through a vacuum, without the mediation of anything else, by and through which their action and force may be conveyed from

one to another, is to me so great an absurdity that I believe no man who has in philosophical matters a competent faculty of thinking can ever fall into it. Gravity must be caused by an agent acting constantly according to certain laws, but whether this agent be material or immaterial I have left to the consideration of my readers.[46]

In fact Newton believed that, far from being a property of matter, gravity was either the spirit of nature (following Henry Moore and the Cambridge Neoplatonists, who were his contemporaries), or else that it was directly the agency of God (which would be more to his liking for theological reasons). How ironic that Newton has been universally proclaimed as the very pillar of the mechanical philosophy!

But the remarkable thing is that this failure to understand what gravity is makes no difference whatsoever to Newton's ability to consider gravity mathematically. He can discover the mathematical law of gravitational force between two bodies without having to know anything at all about the nature of gravity. This is really Newton's major discovery: that it is possible to do what we now call 'physics' (the term wasn't used until the nineteenth century) without having to know the nature of what it is we are dealing with. If it were otherwise, physics as we know it would not have been possible. Newton's way – what Cohen calls the 'Newtonian style' – is to proceed by disciplined imagination to propose mathematical models of physical situations, solve problems in the mathematical model, and translate the solutions back into the original physical situation.[47] This works extremely well, enabling Newton to make discoveries which it would be difficult to imagine being made in any other way. For example, he found mathematically that the gravitational force everywhere inside a hollow shell of matter (i.e. the force due to that shell) is zero if the force obeys an inverse square law. It would be impossible to discover this empirically – though subsequently it might be verified experimentally, as it has been in the case of the electric field, which also obeys an inverse square law. We are now so used to this 'Newtonian style' that we just take it for granted. It is, after all, what we mean by mathematical physics.

Goethe Returns to the Senses

We like to think that the way in which science developed has a quality of necessity about it, in which case the form that science takes must be necessary and not in any way contingent. But what is necessary about the discovery in 1417 of a Latin manuscript written in the first century AD, describing the Greek philosophy of atomism, which then became the basis for the radical transformation of the philosophy of nature leading to the mechanical philosophy and all its ramifications? Surely such a discovery is contingent? It is an example of how a single factor can change a whole situation, but not a case of necessity. Yet looking back now, we tend to endow the way that science developed with a quality of necessity as if it could not have been otherwise. Pointing this out does not imply in any way that science somehow isn't true. Of course it's true. But it's not the only possibility, and for as long as we think it is we will be unable to transform our understanding of our relationship with nature, instead of just tinkering with it at the edges.

The founders of modern science were dedicated to the mathematical approach to nature. What were called the 'primary qualities' were simply those aspects of nature that appeared in the light of mathematics. Although it *is* nature that shows up in this light, this is by no means the only way that nature can appear. As we have seen, the ascendancy of the mathematical was accompanied by the downgrading of the sensory. But there is no necessity here. It is possible for the mathematical aspect of nature to be emphasised without this implying in any way that it is superior to nature as revealed through the senses, or conversely that the sensory is inferior to the mathematical. However, this is just what happened historically: sensory *experience* was relegated to second place in favour of the mathematical.

The influence of the mathematical came in the first place from the Arabs – whom the medieval Europeans referred to as 'our Arab masters'.[48] With the Arabs it seems that mathematics was not cultivated in isolation, but always balanced with other pursuits, such as music and poetry. However, this factor seems to have been left out when mathematics was imported into northern Europe, where as a consequence the emphasis on mathematics became much more one-sided. In the thirteenth century, Roger Bacon said in his *Opus Maius* that mathematics was the 'door and key ... of the sciences and things of this world', and concluded: 'wherefore it is evident that if, in the other sciences, we want to come to certitude without doubt and to truth without error, we must place the foundations of knowledge in mathematics'.[49] It is astonishing how this remark made over eight hundred years ago encapsulates the one-sided mathematical approach that western science has worked with ever since.

This is what Goethe reversed when he returned to the senses and put sensory *experience* first instead of the mathematical. Adopting Roger Bacon's phrase, we could say that for Goethe the senses were the 'door and key' to science. At first this seems unremarkable. After all, this is just what most of us would have assumed anyway – since most of us would probably be unaware of the formative influence of mathematics, and think that science is based *directly* on the evidence of the senses (the philosophy of empiricism). But Goethe does not return to the senses in the empirical sense of relying on the evidence of the senses to gain information about a phenomenon. He was concerned with nature as it comes to presence in the *experience* of the senses. This means putting attention into the sensory experience itself, entering into the lived experience of sensory perception, so that rather than just being 'sensory' in the empirical sense, it is better described as the 'sensuous' experience, or perception, of the phenomenon.[50] Doing this reverses the direction of the automatic learning sequence, and shifts experience away from the verbal-intellectual mode of apprehension into the sensuous-intuitive experience of phenomena.

We tend to rely for the most part on the verbal-intellectual mode of apprehension, because this is what is developed through education in modern western culture. The verbal-intellectual mind functions in terms of abstract generalities that take us away from the richness and diversity of sensory experience – this is both its strength and its weakness. It is focused on what is the same in things, their commonality, so that even without our realising it we become immersed in uniformity and cease to notice differences. For example, if there are two leaves of a tree, as a matter of habit we will tend to see them in a general way as just 'leaves' and overlook the differences between them. This is a consequence of what psychologists call the process of automatisation or habituation. The normal learning sequence goes from the sensory experience of concrete cases to the abstract generalisation. Thus, in the case of the leaves, whereas to begin with we might see each leaf concretely in detail, we eventually replace this with the mental abstraction 'leaf'. When this happens our attention is transferred from the sensory experience to the abstract category, so much so that, without our being aware of it, we begin to experience the category more than we do the concrete instance. When this stage is reached what we 'experience' is only an abstraction triggered by the sensory encounter, and not the concrete case itself. This stage of automatisation, where we experience the category and not the actual occurrence, is demonstrated very clearly in the well-known anomalous playing card experiment.[51]

Goethe's way of thinking goes in the opposite direction to this learning sequence – which, incidentally, is necessary for coping with our daily lives. He redirects attention into the experience of the senses, and in doing so he thereby withdraws it from the verbal-intellectual mind. There is no question here of trying to 'stop' the verbal-intellectual mind which works with abstractions – any attempt to do so would have just the opposite effect. By practising active seeing (which means directing attention into the sensory, instead of just passively experiencing a sense impression), the verbal-intellectual mind is 'suspended', so that attention is brought back into the phenomenon itself instead of being trapped in verbal-intellectual generalities. Goethe puts the phenomenon at the centre of attention and he keeps it there (it's hard work because it reverses the habitual direction of experience.) By redirecting attention into sensuous experience he plunges into the sheer phenomenality of the phenomenon. This reverses the usual direction of the process of habituation from experience to generality, and thereby promotes the process of deautomatisation and hence a renewed encounter with the phenomenon itself.

But this redeployment of attention into sensuous perception by active looking – what could be called reversed seeing – is only the first stage. After this there comes the stage of what Goethe calls 'exact sensorial imagination', and which he describes as 'recreating in the wake of ever-creative nature'. The aim here is to visualise the phenomenon as concretely as possible – not to fantasise about it, embellishing it, but to imagine it as closely as we can to the phenomenon we encountered through sense experience. This is an exacting discipline, trying not to add anything which is not there in the phenomenon, and at the same time not to leave anything out. Here again the phenomenon itself is made the focus of our attention. But whilst focusing on the phenomenon in this way, what we are doing effectively is to make the phenomenon more 'inward'. We are going into the phenomenon, as we do in active looking, but now we are going into it by bringing it into ourselves.

This means that we are creating a 'space' for the phenomenon by means of our attention so that we can receive it instead of trying to grasp it – so that we become participant in the phenomenon instead of an onlooker who is separate from it. If we now return to the sensory encounter with the phenomenon, we will find that our senses are enhanced and we begin to become aware of the more subtle qualities of the phenomenon. As we follow this practice of living into the phenomenon, we find that it begins to live in us. Whereas the intellectual mind can only bring us into contact with what is finished already, the senses – enhanced by exact *sensorial* imagination – bring us into contact with what is living, so that we begin to experience the phenomenon dynamically in its coming into being.

This is exemplified by Goethe's way of seeing the colours that appear when we look through a prism. Since the colours only appear wherever there is a visual boundary, a simple way of doing this is to construct a straight black/white boundary and look at it through a prism – the boundary and the axis of the prism should both be horizontal for the optimal effect. Vivid colours are seen at the boundary, and which they are depends on its orientation. If black is above white the colours seen are red, orange and yellow; if white is above black the colours are pale blue, a deeper blue (sometimes called indigo), and violet. As soon as we label them we begin to think of them as separate colours. But they are not so clearly distinguished in sensuous experience, where we find they seem to merge one into the other as we move through them with our eyes. When we put attention into seeing, as if we were going into the colours through our eyes, we become aware of the sensuous quality of each colour – for example, the redness of red, that red is *red*. We do not usually experience this sensuous quality, but just register the colour as 'red' or 'blue', and so on, by observation – i.e. by sense perception which gives us the information that it is 'red' but does not take us into the experience of red.

The second stage is the practice of exact sensorial imagination. Now we put aside the physical manifestation and work entirely in imagination, trying to visualise what we have seen as exactly as we can. As we move through the colours at a boundary in imagination, we begin to experience their sensuous quality as if we were within the colours – one student described this as feeling like she was swimming through the colours. We find there is a dynamic quality in the colours at each boundary. What we experience is not separate colours – red, orange, yellow, or pale blue, deeper blue, violet – but something more like 'red–lightening–to–orange–lightening–to–yellow' as a dynamic whole, and similarly with the darkening of blue to violet. There is a sense that the colours are different dynamic conditions of 'one' colour. This dynamic quality gives us an intuition of the wholeness of the colours at each boundary. This is not given directly to sense perception, but appears when sensuous perception is sublimed into intuition through the work of exact sensory imagination. In this way the sensuous-intuitive mode of perception replaces the verbal-intellectual mode. The colours are no longer thought of as being separate (verbal-intellectual) but are experienced as *belonging* together (sensuous-intuitive). The way to the wholeness of the phenomenon is through the doorway of the senses and not the intellectual mind.[52] We find there is the sense of a *necessary* connection between the *qualities* of the colours at each boundary. It is not just accidental, for example, that the order of the colours is red, orange, yellow – and not red, yellow, orange – but is intrinsic to the colours themselves. This kind of connection between the qualities of the colours is missing from the Newtonian theory which asserts that light consists of colours which are separated when it is passed through a prism. In this case there is no intrinsic necessity in the order of the colours, but only an order that is imposed extrinsically by the attribution of a wavelength to each colour.[53]

The transition from the abstract verbal-intellectual mode of apprehension to the concrete sensuous-intuitive mode, is exemplified very clearly in Goethe's account of metamorphosis in the life of the plant. Recent work in developmental genetics has thoroughly vindicated Goethe's insight using the techniques of modern research.[54] However, what matters most with Goethe is not so much the *fact* of metamorphosis as experiencing the metamorphic *way of seeing* – and this is the factor that is missing in the research laboratory. Here again, Goethe's way proceeds by active looking and exact sensorial imagination. We can see this most readily by considering the leaves up the stem of the flowering plant. We begin by focusing attention closely on the unique particularity of each leaf, looking carefully at its form and structure, and then trying to visualise it as well as we can. When we look at it again we will find that our perception is enlivened. Now when we follow the same procedure with the next leaf, we will notice differences, and yet at the same time there is a sense of similarity to the first leaf. After repeating this process with several leaves as we move up the stem, we can go on to practise the exact sensorial imagination of the sequence. We visualise the first leaf, and then move in imagination to the next leaf, and so on. We will soon begin to have an intuition of the sequence as a movement that is a dynamic whole – a dynamic gestalt – instead of just a series of steps.

We begin to have the intuition that we are seeing 'one' leaf manifesting in different forms. We have the sense that this 'one' leaf is intrinsically dynamic, and that this dynamic whole is a movement of *differencing* which produces 'multiplicity in unity'. The verbal-intellectual mind, in contrast, focuses on the sameness of the different leaves, and from this abstracts the notion of 'one' leaf as simply what all the leaves have in common – their lowest common denominator. All differences are excluded from this 'one', whereas for the sensuous-intuitive mode of perception the differences are within the 'one'. Instead of abstracting unity from diversity, we have the intuition that the diversity is within unity. This becomes clear when we work concretely with the plant in the way that Goethe indicated. If we don't do this, and instead just follow our usual proclivity for abstract thinking, we will fail to distinguish between these two different modes of unity, and fall back into the mental attitude of an onlooker, i.e. thinking of the plant in its finished state, instead of participating in the coming-into-being of the plant in our thinking – what Craig Holdrege calls 'learning to think like the plant lives'. The key thing is that, where the verbal-intellectual mind sees 'sameness in the midst of difference', the sensuous-intuitive mind sees 'difference in the midst of sameness'. There is a reversal of perception here that it is hard to convey unless it is experienced – it's as if our perception of unity and diversity is turned inside out, so that diversity is seen *within* unity instead of unity being abstracted *from* diversity. To do this we have to turn it round and experience the unity from the 'point of view' of the living plant which is bringing forth multiplicity out of itself, instead of from the point of view of an observer who is trying to find unity in a multiplicity which is already given. This is an example of the difference to which Heidegger refers when he says 'the way in which an entity we are interpreting is to be conceived can be drawn from the entity itself, or the interpretation can force the entity into concepts to which it is opposed in its manner of Being'.[55]

So far we have only considered metamorphosis in the leaves of the flowering plant. But in *The Metamorphosis of Plants*, Goethe is concerned with all the organs of the plant – sepal, petal, stamen, pistil – which he sees as modifications of one organ.[56] He describes metamorphosis as the 'process by which one and the same organ presents itself to us in manifold forms', and in a letter to Herder he describes this 'one' organ as 'the true Proteus... who can conceal and reveal himself in all forms' – Proteus being the Greek God who can present himself in manifold forms, always differently, and yet always Proteus. The movement of thinking here is indeed very different from looking for uniformities and commonalities in order to find a 'general plan common to all organs', which is the approach so often wrongly attributed to Goethe. The dynamic idea of the unity of nature that we find in Goethe is also very different from the kind of unity we find in the universal laws of nature, which came from the mathematical approach in science, and which had such a cultural impact in the Enlightenment. The unity of this universal also leads our thinking in a direction that excludes difference – and eventually degenerates into uniformity – whereas the dynamic unity we find in life leads us to recognise diversity as creative unity.

There are often situations in which we can learn to recognise the difference between seeing 'unity in diversity' or 'diversity in unity'. A few years ago I visited the Horniman Museum in South London to see the new aquarium that had just been installed. Afterwards I wandered through to the anthropological exhibits, where I found myself in one section standing in front of a large glass case extending the entire length of the wall, containing masks and other head gear, decorated shields and weapons of various kinds – all the shields were grouped together, and similarly the other artefacts – arranged in a way that gave a sense of their belonging together. No attempt was made to relate them to each other explicitly – it was just the way they were arranged. In the case of the decorated shields, for example, they were arranged in a series, so that the eye could move along from one to another whilst at the same time taking in the series as a whole. I was reminded of the way that Goethe laid out the leaves of a plant in a series, and I realised that here also with these human artefacts there are two ways of seeing. In one way we can see that they are all based on the same plan, and that this common plan is the unity in the diversity. The movement of thinking here is *away* from difference *towards* unity. But in this movement, as difference is left behind, the unity begins to appear as a reduction of the diversity of the phenomenon. It becomes fixed and abstract, and there is the feeling that it lacks something as the differences recede into the background, leaving what is the same standing out more clearly. This is the kind of unity we find when we begin 'downstream' with the finished products, as we must, but then go even further downstream to abstract unity from their diversity. But there is another way of seeing, which also begins with the finished products, but moves in the opposite direction and goes back 'upstream', by placing ourselves within the coming-into-being of diversity. When we do this we see the unity concretely as a *productive* unity. We are now 'on the other side', no longer an onlooker standing outside of what we see, but as if we ourselves were within the productivity, participant in the producing instead of standing in front of the products. The unity can therefore no longer be abstract, but includes difference within it as a natural consequence of the productivity. Difference stands out now, instead of receding into the background, but the difference is now the *dynamic* unity of the productivity. In other words, the unity is generated in the very act which differences, instead of being abstracted by ignoring the differences. As I stood in front of the decorated shields in that glass case, I found that I could practise going from one way of seeing to the other – from unity in diversity (the finished products) to diversity in unity (the productivity). It was evident in this experience that diversity *is* dynamic unity. So when we see diversity we are looking at unity, but not recognising it at first – and so we go looking for it in another direction, away from the phenomenon into abstraction.

Goethe and the Bimodal Brain

The difference between the verbal-intellectual and the sensuous-intuitive modes of experience is correlated with the difference between the left and right hemispheres of the brain. This is not in any way intended to imply neurological reductionism. Although the discovery of the hemispheric differentiation of functions became very popular in the 1970s, the tendency then was to divide human functions into two separate lists, allocating each function to one side of the brain or the other. This led to many ridiculous exaggerations, most notably the one which effectually portrayed the left hemisphere as 'snaps and snails and puppy dogs' tails' – which was identified as being male – and the right hemisphere as 'sugar and spice and all things nice', and which of course was female. It is little wonder that 'the subject of hemisphere differences has a poor track record, discouraging to those who wish to be sure that they are not going to make fools of themselves in the long run.'[57]

But this has now changed, and it has become possible to take it seriously again, especially since the publication of McGilchrist's magnum opus, *The Master and His Emissary*, from which the following account is taken.

The most fundamental difference between the hemispheres lies in the way they attend to the world:

> One of the more durable generalizations about the hemispheres has been the finding that the left hemisphere tends to deal more with pieces of information in isolation, and the right hemisphere with the entity as a whole, the so-called *Gestalt*.[58]
>
> Then there is the *primacy of wholeness:* the right hemisphere deals with the world before separation, division, analysis has transformed it into something else, before the left hemisphere has *re*-presented it. It is not that the right hemisphere connects – because what it reveals was never separated; it does not synthesise – what was never broken down into parts; it does not integrate – what was never less than whole.[59]

But the key difference which emerges is that the right hemisphere is concerned with the immediacy of lived experience – 'the right hemisphere delivers what is new as it "presences"'(p.179) – whereas the left hemisphere is concerned with the representation of experience – it 're-presents' what is 'present' to the right hemisphere. Because we only *know* things when they are represented, there is a tendency for us to rely on the world as it appears through the left hemisphere, and therefore to overlook the primacy of experience, and indeed to mistake the secondary representation of experience for the experience itself – which is very familiar in phenomenology (the light which the discovery of hemispheric difference can throw on phenomenology, and reciprocally the way in which phenomenology can illuminate the world as experienced through the two hemispheres, is potentially a valuable insight to be explored).

Another key difference is that 'where the left hemisphere is more concerned with abstract categories and types, the right hemisphere is more concerned with the uniqueness and individuality of each existing thing or being' (p.51). Not surprisingly, therefore, since it 'attends to individual things in all their concrete particularity' (p.153), it is the right hemisphere which mediates the experience of the senses, whereas the left hemisphere mediates the verbal-intellectual representation of experience. Goethe's concrete way of working by returning attention to the senses and withdrawing it from the verbal-intellectual mind, therefore promotes a shift from the dominant (but not primary) left hemisphere back to the right hemisphere; from what is known and familiar to what is living and new; from what is re-presented to what is 'present' – 'the senses are crucial to the "presence" of being' (p.153).

We can now see the neuropsychological correlate of the difference between the verbal-intellectual and the sensuous-intuitive modes of experience. We can see that Goethe's way of working, by returning to the senses through active seeing and exact sensorial imagination, brings about a shift from the left hemisphere dominance of the verbal-intellectual mind to the right hemisphere experience of the wholeness of what is livingly present which is characteristic of the sensuous-intuitive mind. This may well be Goethe's greatest discovery: how to encounter what is active and living in nature by means of the senses and their enhancement, instead of remaining in contact only with what is already finished by relying on the intellectual mind. What we can now add to this is the discovery of the neuropsychological correlation between Goethe's way of science and the difference between the modes of functioning of the two hemispheres of the brain. Perhaps such a contemporary approach could provide a doorway through which Goethe's sensuous-intuitive way of science can be introduced into the world today.

3. Goethe and the Dynamic Unity of Nature

Whenever Goethe is mentioned in connection with science, it is usually in the context of his work on colour, and especially his disagreement with Newton. Consequently his approach to science is presented from the outset as being controversial. But this tendency to focus on Goethe's more controversial work has the unfortunate consequence of drawing attention away from his other, equally important work on metamorphosis in plants. Although this work was quite radical at the time, it is certainly not controversial. It is in fact precisely what modern biology has discovered in its own way. What Goethe said about metamorphosis is confirmed today by developmental genetics.[1] The puzzling thing is, as one professor of genetics put it to me, how Goethe could have got it so right over two hundred years ago without the resources of modern genetics. The answer is that he did it by learning 'to think like a plant lives' through the practice of active seeing and exact sensorial imagination.[2]

The Idea of Metamorphosis

Goethe begins *The Metamorphosis of Plants* (1790) with the observation that:

> Anyone who observes even a little the growth of plants will easily discover that certain of their external parts sometimes undergo a change and assume, either entirely, or in a greater or lesser degree, the form of the parts adjacent to them.[3]

By 'external parts' he means the various organs growing from the stem of the plant. Firstly there are the vegetative leaves winding up the stem,

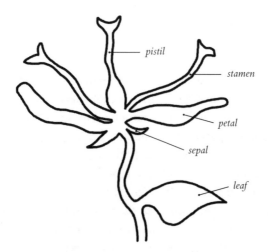

Figure 2. The parts of the plant.

and then the rings of organs comprising the flower: the sepals which contain the floral bud, and which open to reveal one or more rings of petals surrounding an inner ring(s) of stamens, all of which surround the central organs (pistil and ovary) at the end of the stem where reproduction takes place and seeds are formed (see Figure 2).

Goethe brings out what he means more clearly in his next observation:

> So the simple flower, for example, often changes into a
> double one, if petals develop in the place of stamens and
> anthers. These petals may either perfectly resemble the other
> petals of the corolla, both in form and colour, or they may
> still retain visible signs of their origin.

An example of this is provided by the difference between the wild and the cultivated rose. The wild rose has a widely open flower with a single ring of petals, within which there are several rings of stamens. The cultivated rose, on the other hand, has a closed flower consisting of several rings of petals, within which there is a single ring of stamens. The difference in appearance is striking: on the one hand a simple flower opens to view, and on the other an enclosed flower which hides itself and has become a symbol of beauty and mystery. The difference botanically is that rings of stamens in the wild rose have 'metamorphosed' into rings of petals. So where stamens should be, now there are petals – an example of what Goethe calls retrogressive metamorphosis, because here the plant takes a backward step with respect to its normal developmental sequence.

When we notice the fact that petals sometimes appear in the place of stamens, we may have the intuition that there is some kind of inner connection between petals and stamens. Organs which appear at first to be distinct and separate, now seem to belong together. But are there instances in the normal developmental sequence of the plant where we can recognise this 'secret affinity', as Goethe puts it, between petals and stamens? There are indeed. It is so evident in the white water lily, for example, that, when we recognise it, we could easily believe the idea had become 'visible' and that we are seeing it with our eyes. In this plant we find several intermediate stages between petals and stamens. Here again there are several successive rings of organs, with each ring showing a distinct intermediate form on the way from petal to stamen (see Figure 3).[4]

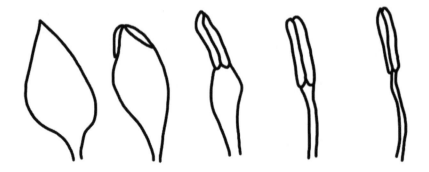

Figure 3. Intermediate stages between petal and stamen.

Several developmental stages can be seen simultaneously here, so that when we look at a waterlily the overall effect is that we seem to 'see' one organ turning gradually into another one. But this is not what is happening: a petal does not materially turn into a stamen. Rather, what we are seeing here is one organ manifesting in different forms, and not one organ turning into another one – i.e. no finished petal changes into a stamen. The metamorphosis is in the embryonic stage of plant growth and not at the adult stage (see Figure 4).

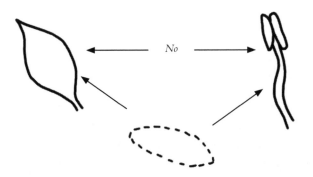

Figure 4. No metamorphosis from the adult form.

Goethe expresses this as follows (referring specifically to the retrogressive case of petals in the place of stamens):

> If we see that in this way it is possible for the plant to make
> a retrograde step and reverse the order of growth, we shall
> become all the more aware of the normal course of Nature,
> and shall learn to understand those laws of transformation by
> which she produces one part out of another and creates the
> most varied forms by the modification of one single organ.

Almost the first thing we notice here is the possible misunderstanding which this invites. Goethe begins by referring to the transformation that 'produces one part out of another', which in the context could lead us to think that a stamen is produced out of a petal at the adult stage instead of in the embryonic growth form of the organ. But he then goes on immediately to say that nature 'creates the most varied forms by the modification of one single organ', which expresses the idea very clearly. All the organs appended to the stem of the plant are to be seen in this way – from the stem leaves through to the reproductive heart of the flower.[5] So in the next paragraph Goethe considers, not just adjacent organs like petals and stamens, but all the organs as 'modifications of one single organ':

> The secret affinity between the various parts of the plants
> such as leaves, calyx [sepals], corolla [petals], and stamens,
> which are developed one after the other and as it were one

out of the other, has long been recognised in a general way by naturalists; indeed much attention has been given to the study of it. The process by which one and the same organ presents itself to us in manifold forms has been called *the metamorphosis of plants.*

Here again, the first thing we notice is the suggestion that the successive organs are developed 'one out of the other', which could be misleading if we took it to mean that an organ at the adult stage transformed physically into another one. Goethe certainly does not mean to say this, as we can see from the fact that he adds 'as it were' to qualify it. Immediately after this comes the clear statement that metamorphosis is 'the process by which one and the same organ presents itself to us in manifold forms', which is completely different from the idea of one organ turning into another one. It is the ability of the vegetative shoot to develop into different forms which leads to the diversity of organs, and not some miraculous ability on the part of a finished organ to change its form into a different organ. The metamorphosis is in the earlier embryonic stage of the coming-into-being of the organs, and not at the later adult stage of organs that are already finished. Goethe's way of thinking is intrinsically dynamic: it goes back 'upstream' into the coming-into-being of the organs, instead of beginning 'downstream' with the organs that are already formed. Metamorphosis is only to be found in the coming-into-being, and the failure to realise this leads us to look in the wrong direction by trying to understand metamorphosis in a downstream way. This is the source of much of the misunderstanding about Goethe's work.

We are now much more familiar with this dynamical thinking as a consequence of the way that biology has developed, so that it is much easier to understand the idea of metamorphosis in the right way than it might have been previously. For example, stem cell research has drawn our attention to cells in embryos (embryonic stem cells) which are unspecialised – they proliferate by cellular division without specialising – but which can also differentiate, in which case a stem cell can take on any one of a number of specialist functions, for instance, becoming a liver cell, or blood cell, or nerve cell, and so on. This is an astonishing discovery, and yet it is clearly in line with the idea of metamorphosis in the plant, where the vegetative cone of the plant remains in an embryonic state and can therefore develop into different organs according to circumstances. Similarly, a stem cell can develop into a specialised cell in different ways according to circumstances – but we would never expect to see an already formed blood cell turn into a liver cell, for example. The fundamental process of life which Goethe recognised in the metamorphosis of the plant, is dynamically similar to the embryonic development of specialised cells from the stem cells in the growth of organisms.

Protean Thinking

So far we have seen that in some sense the different organs of the plant are one organ. But what kind of 'one' is this? What kind of 'one' can present itself in manifold forms, and what is the relationship between this 'one' and the 'many' forms in which it manifests? We are now going to explore this question in some depth.

In his own time, and indeed ever since, an answer seems to have been given to this question which is based on an assumption that is in fact nowhere to be found in Goethe's work. This is the assumption that he was searching for what all the different plant organs have in common – their 'lowest common denominator'. By trying to find what is the same in all of them, i.e. in which there is no difference at all between them, it is supposed that Goethe discovered a unity in the diversity of the organs. The movement of thinking in this case evidently has the effect of *excluding* difference from unity. We can see this clearly in some of the statements that have been made about what Goethe was doing – for instance, that he 'was transfixed by uniformities and commonalities in nature', and that he sought for 'the general plan common to all organs' by trying to find 'the simplest form of plant organ from which the anatomist's mind had stripped all the specializations required by the organs of real living plants'. Statements such as these, which are typical, clearly do not portray nature in the way that Goethe expressed to Schiller, as 'working and alive, striving out of the whole into the parts'. On the contrary, they portray nature more as dead and finished.

When we read what Goethe says carefully, paying attention to the movement of thinking, we can see for ourselves that he was doing something radically different from just looking for what all the plant organs have in common. We have already seen that he says 'nature creates the most varied forms by the modification of one single organ,' and describes metamorphosis as ' the process by which one and the same organ presents itself to us in manifold forms'. Elsewhere, in letters and the diary of his Italian journey, he says that he is 'becoming aware of the form with which again and again nature plays, and in playing brings forth manifold life', and that 'the thought becomes more and more living that it may be possible out of one form to develop all plant forms'. Notice that he does not say the form with which nature plays again and again is nature's model or ground plan of the plant, just as he does not say that he is trying to reduce all plant organs to one form.[6] Yet again, on another occasion when referring to the organs of the plant, he says: 'It had occurred to me that in the organ of the plant which we ordinarily designate as *leaf*, the true Proteus is hidden, who can conceal and reveal himself in all forms'. Reading what Goethe says, it is difficult not to get the sense that he is doing the very opposite of searching for what all the different organs have in common. He is talking about the creation of difference within unity, not arriving at unity by the exclusion of difference. The direction of his thinking is the other way round.

The reference to Proteus gives us an indication of the direction Goethe's thinking takes. Proteus is the Greek God who can hide and reveal himself in any form he chooses. He can present himself in manifold forms, ever differently, and yet it is always Proteus. Now we would not try to understand Proteus by collecting together different manifestations and trying to see what they all have in common. Such a procedure would be far 'too late'. What is essential about Proteus is the coming-into-being, the *appearing,* and not the specific form in which he appears. The attempt to find a common identity based on the different appearances could only result in an 'average Proteus,' which is an absurd notion that would only take us even further away from the ever-dynamic Proteus. So clearly, Goethe does not want us to look at the organs of the plant and find what they have in common, excluding all the ways in which they are different from one another and including only the ways in which they are the same, until at last we arrive at a kind of 'average organ' which is the common plan according to which they are all formed. It takes only a moment's thought to realise that no real differences could ever be produced from such an 'average organ,' because it is reached by excluding all differences in the first place. It is a cul-de-sac.

So Goethe is not saying: begin with the finished organs as they are on the adult plant and then try to abstract a unity from them. If this were the case we could *only* end up with what they all have in common. For Goethe the organs in their finished state are already 'downstream', and to abstract from them only the unity of what they have in common is to go even further 'downstream'. But Goethe goes in the opposite direction and tries to catch nature 'in the act' – i.e. 'working and alive, striving out of the whole into the parts'. He goes back 'upstream' from the organs in their finished state, so that he doesn't derive the unity *from* the diversity, instead he 'brings the diversity back into the unity from which it originally went forth'.[7] In this way the movement of his thinking can follow the coming-into-being of the organs and end with them in their finished state (see Figure 5).

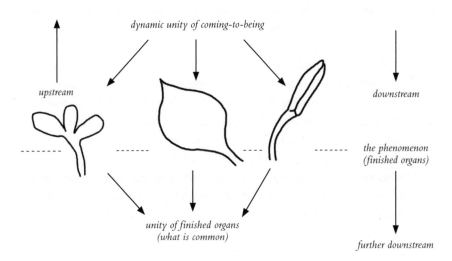

Figure 5. The coming-into-being of the plant.

We always begin with the phenomenon, which is already downstream. The difference is whether we then go further downstream in our search for unity, or whether we go upstream into the coming-into-being of the phenomenon. If we go upstream we discover the dynamic unity of the emerging organs, so that we now come into the phenomenon from the unity instead of trying to come to the unity from the finished phenomenon. So we can begin to think in a Protean way that 'creates the most varied forms by the modification of one single organ'. If we don't recognise the difference between these two movements of thinking, then we easily fall into the error of trying to 'reach the milk by way of the cheese' by projecting the unity abstracted from the finished organs back into the beginning, as if this unity of the dead end were the unity of the living origin.

Goethe and Schelling

Goethe came to a fuller understanding of the dynamical form of his own thinking partly as a result of his conversations with Schelling – the highly precocious philosopher who was appointed to the chair of philosophy at the University of Jena in 1798 as a result of Goethe's influence. It was through these discussions that Goethe came to realise that the direction of his own thinking took him upstream from seeing nature as product into seeing nature as producing. Schelling emphasised that to understand nature we must rise from nature as fact to nature as 'the action itself in its acting'. He says that: 'In the usual view, the original productivity of nature disappears behind the product. For us the product must disappear behind the productivity'.[8] Of course, in emphasising the reversal Schelling in no way intends to imply a separation between the productivity and the product – his thinking here is dynamical. We can now easily recognise that this is indeed what Goethe is doing by going back upstream into the coming into being of the phenomenon, instead of beginning downstream with the phenomenon in its finished state and trying to explain it from there – which can only take us even further downstream. The difference between nature as productivity and nature as product is often described in terms of the distinction between *natura naturans* ('nature naturing') and *natura naturata* ('nature natured'). It is the latter which we think of as the natural world, whereas for Schelling it is really the former that is implied by the idea of *Nature*. Goethe's remark that it is possible to present nature as 'working and alive, striving out of the whole into the parts', clearly refers to nature in the sense of 'nature naturing'. The distinction between *natura naturans* and *natura naturata* is often attributed to Spinoza, whose work was read and admired by Goethe, Schelling, and others at the time. However, this distinction did not in fact originate with Spinoza, for whom it was certainly important, but will be found earlier in Renaissance nature philosophy, and goes back to Scotus Eriugena in the tenth century.

The Self-Differencing Organ

If one and the same organ presents itself to us in different forms, then each organ *is* that organ, but differently, and not *another* organ – Proteus is always one and the same Proteus, but differently, and not another Proteus. It is always the very same one and not another one, and yet it is always becoming different from itself. It becomes other without becoming another – the other of itself and not another one. Goethe's 'one and the same organ' manifesting as different forms is a self-differencing organ producing differences of itself. *So the different organs we see are the self-differences of one organ.* What we discover here is the extraordinary idea of self-difference instead of self-sameness, the idea that something can become different from itself whilst remaining itself instead of becoming something else. When we go upstream into the coming-into-being we discover the self-differencing organ which appears downstream as several different organs – to borrow from Gilles Deleuze, we find 'there is *other* without there being *several*'[9] So we find that the unity of coming-into-being is the dynamic unity of self-differencing, in which difference is intrinsic to unity. Here the unity is the very dynamics of self-differencing. There is no separation here (if we find it in our thinking, it is because we have 'fallen downstream' without noticing): the self-differencing *is* the unity and concomitantly the unity *is* the self-differencing. This dynamic unity is evidently the very opposite of the unity of the finished products, which is the static unity of self-sameness that is reached by the *exclusion* of difference (see Figure 6).

unity of coming-into-being	=	dynamic unity of self-differencing
unity of finished products	=	static unity of self-sameness

Figure 6. Modes of unity.

When one thing is different from another thing, the distinction is extensive; but when something is different from itself the distinction is intensive.[10] What this means will become clearer with the example provided by the hologram. If we have a transmission hologram on a glass plate – let us say of a horse galloping towards us – what would we expect to find if we divided the plate physically into two halves?

If we had a photographic plate instead of a hologram, we know what the answer would be: two halves of the plate with half the horse on one and the other half on the other. What is surprising about hologram division is that we would find the *whole* horse on each of the two halves of the plate.[11] We can divide the photographic plate but not the hologram of the horse. The contrast with a photograph is striking: if we want another photograph we have to make a copy of the first one, and then there will be two photographs – one and another one. But there cannot be 'two' holograms here – even though it looks like there are physically – because this does not take into account the optical indivisibility of the hologram, whereby the attempt to divide it results in the whole again instead of two halves. If we now ask how many there are, what can we say? We cannot say there are two, because this would be no different from the case of photographic reproduction, where there clearly are two (one and another one). In the case of the hologram it seems that each is the *same* one – one and the other of itself instead of one and another one. In a sense there is only one and not two, yet clearly not in a numerical sense because then we would be unable to distinguish this case from the original hologram before it was divided. The division of the hologram is *intensive* because it remains whole when divided, and consequently the distinction between the 'two' which are one and the very same one (one and the other of itself) is an intensive distinction. We can call this 'multiplicity in unity'. The division of the photograph, on the other hand, is an *extensive* division because it results in two halves. Copying the photograph is also extensive, because the result is 'one and another one' and not 'one and the other of itself'.

This difference can also be seen in the vegetative reproduction of plants. When a gardener propagates a plant by taking cuttings, what is happening *organically* is similar to what is happening *optically* when a hologram is divided in the manner described above. For example, if a leaf is taken from a fuchsia plant, and divided into several pieces, each of which is then planted separately, eventually these cuttings will grown into adult fuchsia plants. So where we had a single plant to begin with, there will now be several plants, which can be separated and moved to different locations as if they were simply physical objects. But *organically* there is only One plant here, because each is the very same one and not another one. Like the hologram, the plant is divisible and yet remains whole – so that it is really 'indivisible' in a subtle sense. Organically each one is the very same plant, so where there appear to be many plants there is really One plant which is prior to separation. The plant has become multiple without becoming many plants – even though this is how it seems to us when we count the plants, because when we do we count them as physical objects. The difference is between the non-numerical multiplicity of 'multiplicity in unity' and the numerical multiplicity of many ones. This must be an *intensive* multiplicity otherwise the unity would fragment.

So here we have a unity which includes multiplicity *within* it without being divided – and thereby ceasing to be unity. Extensively we can have either one or many – one only or many ones – but intensively we can have one and many at the very same time because the one *is* many. The difference here is that the one can be multiple intensively without being many extensively. Such a 'multiplicity in unity' constitutes an intensive dimension of One, as distinct from the extensive dimension of many ones. In what follows we shall use an initial capital letter in this way to donate the 'one'which is intensively 'multiplicity in unity', and a lower case letter to donate the 'one' which is extensively one of 'many ones'. Thus, in vegetative propagation there is One plant organically where we see many plants. So the unity of coming-into-being, which is the dynamical unity of self-differencing, produces 'multiplicity in unity' which is an intensive dimension of One.

We can see this quite strikingly in other examples of vegetative reproduction, where what appears to us as many plants is in fact One plant be-ing itself multiply. Consider a strawberry bed in the garden, for example. We see this as many strawberry plants, whereas it is in fact One plant. The strawberry propagates by sending out a creeping runner along the ground, from the tip of which it puts out roots and a new strawberry plant shoots out. But organically this is the very same plant – the other of itself and not another one. This means that the entire strawberry bed is organically One plant, a 'multiplicity in unity' instead of many separate plants, in which 'there is *other* without there being *several*'. Yet another very striking example of the difference between an intensive and an extensive distinction is provided by the growth of potatoes. John Seymour describes this:

> The potato is not grown commercially from seed, but from sets, which are just potatoes, and so all the potatoes of one variety in the world are *one plant*. They are one individual that has just been divided and divided. [12]

To produce a new variety it is first necessary to fertilise a plant with the pollen of another. After that:

> ... The breeder arranges for the new variety to be multiplied by setting the actual potatoes from it – and if it proves a popular variety the original half dozen or so potatoes on the first-ever plant of that variety may turn – by division and subdivision – into billions and billions of potatoes – all actually parts of that first plant. It would be interesting to know how many billion tons that first King Edward plant has developed into during its life![13]

So the King Edward potato is dynamically One plant in space and through time. It is a non-numerical, organic 'multiplicity in unity' which is an intensive dimension of One, and not the numerical multiplicity of many ones we see when we are buying potatoes. We can of course see it both ways: as One which is intensively multiple, or as many which are extensively separate. But in the latter mode we lose the organic 'indivisibility' of the whole, and we see the potatoes as no more than physical bodies like a pile of bricks. Here again we can see the difference between following the coming-into-being and starting with the already finished products. Whereas 'downstream' we see many plants, 'upstream' we discover One plant be-ing itself multiply.

It is this idea of an *intensive* distinction which we need in order to see the transformation of the idea of 'the one and the many' in Goethe's dynamical thinking. However, the examples we have given above to illustrate this kind of distinction only consider multiplicity and not genuine diversity. Hologram division, or plant cuttings, result in identical holograms, or plants, whereas the 'one organ' Goethe is describing can present itself in various different forms – as vegetative leaf, sepal, petal, stamen. Nevertheless, the same form of thinking is needed here for genuine diversity as for simple multiplicity. Here again it is possible to give examples which can function as 'templates for thinking'.[14] For instance, the hologram model can be extended to the case of the multiple hologram, in which several different images can be recorded on one and the same hologram without becoming confused – as they would be in the case of a photograph which had been multiply exposed. The important point here is that each different image is recorded on the entire hologram – not one image on one part of the hologram, and another on another part, and so on. If each exposure is taken at a slightly different angle, and then the angle at which the hologram is looked at is also changed slightly, what is seen is one image (for instance, a horse) turning into a different one (for instance, a cow) in the very same place. This illustrates how the dimension of wholeness can contain many within it in such a way that each one *is* the whole, but differently, because in this case each different image in the hologram is the whole hologram and not part of it in an extensive sense.

A further illustration of the intensive dimension of self-difference is provided by the experience of seeing a multi-perspectival figure, such as the reversing cube or the duck/rabbit. Thus, in the case of the duck/rabbit, each different figure that we see – whether duck or rabbit – is the whole figure. So it is an instance of 'multiplicity in unity'. One figure does not occupy only part of the picture, while the other figure occupies the other part – it is duck/rabbit, not duck *and* rabbit (see Figure 7).

Figure 7. The ambiguous duck/rabbit image.

There are no lines left over, unused, by either figure – and no extra lines need to be added in either case. Each figure is complete in itself, and yet it is not the only possibility. The duck and the rabbit are nested intensively in one another. Either can come into manifestation, but not both together, side by side, extensively – if we try to do this, the duck and the rabbit will each be a duck/rabbit. Each one is the very same One and not another one, but differently. So this illustrates the idea of the intensive unity of self-difference, which includes difference without fragmenting the unity. Here only two figures are included, and we do not know of multivalent figures which are more than bivalent. However, although we may not be able to draw such a possibility, we can certainly conceive it, and thereby consider the possibility in principle of a multivalent figure with three, four, five, ... figures nested intensively. Such a possibility would go some way towards illustrating the intensive unity of the self-differences of the One organ which manifests as leaf, sepal, petal, stamen, and so on – although its limitation is that it is not intrinsically dynamic.

It has already been mentioned that Goethe's way of seeing the metamorphosis of the plant has been confirmed by modern research in developmental genetics. The electron micrograph below shows the development of a floral bud at an early embryonic stage.[15] Here we can see the self-differencing organ coming-into-being as sepal, petal, and stamen. This is Goethe's 'diversely metamorphosed organ' (see Figure 8).

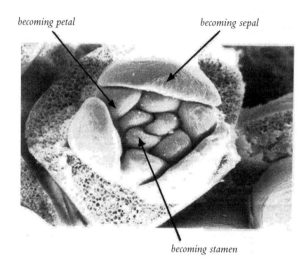

becoming petal

becoming sepal

becoming stamen

Figure 8. Electron micrograph of an embryonic floral bud.

Becoming Other in Order to Remain Itself

Ron Brady, a philosopher who contributed much towards understanding Goethe's morphology, describes the intrinsically dynamic form of life as 'becoming other in order to remain itself'.[16] He says that:

> The forms of life are not 'finished work' but always forms *becoming*, and their 'potency to be otherwise' is an immediate aspect of their internal constitution ... The *becoming* that belongs to this constitution is not a process that finishes

when it reaches a certain goal but a condition of existence – a necessity to change in order to remain the same.[17]

Here we have a clear recognition of the self-differencing organ, ever changing into other modes of itself, so that what we see as the diversity of organs *is* the living unity of the plant.

So far we have considered only the organs of the plant, but we can now expand our horizon to consider the dynamics of 'becoming other in order to remain itself' in the variety of different forms which an individual plant species can take. If we consider a single species of plant, we will see individual plants of this species taking on different forms according to the conditions of the environment in which they are growing. Changes in environmental factors such as soil conditions, weather patterns, the light, and so on, are seen to result in marked differences in the external form (the phenotype) of individual plants of the species.[18] When Goethe travelled across the Alps into Italy, he saw many plants which were familiar to him in Southern Germany, but modified in accordance with the change in environment. Thus, in the Alps he noticed that, in general, branches and stems were more delicate, buds further apart, and leaves narrower, than they were in the same species in Germany. He recognised that in such cases he was seeing different manifestations of the same plant and not different plants. This phenotypic variety of a species is not extensively many different plants, but intensively One plant coming-into-being as different expressions of itself, be-ing itself differently in changing circumstances. The idea of 'the one and the many' is turned inside out here: Goethe does not see many different plants which are basically the same (downstream, static) but One plant be-ing itself multiply (upstream, dynamic). The one is not separate from the many in this way of thinking. On the contrary, what we find here is that, in the words of Gilles Deleuze: Multiplicity is the inseparable manifestation, essential transformation and constant symptom of unity. Multiplicity is the affirmation of unity; becoming is the affirmation of being.[19]

We must be careful when we are trying to conceive the plant in its environment in an organic way that we do not inadvertently fall into a way of thinking that is more appropriate to the inorganic realm. It is all too easy for our thinking to lose sight of the very quality of livingness which is the organism's own 'potency to be otherwise', and for us to fall into thinking of the organism as if it were responding in a mechanical manner to the influences of the environment. The *living* organism does not just adapt to external circumstances in a passive manner, as it would do if it were only an inert body responding to external forces according to the laws of physics. The specific form which a plant takes in its surroundings is not the result of external conditions acting directly on the plant to cause the modification which we observe. The conditions clearly do influence the specific form which the particular plant takes, but they do not cause it. Such a way of thinking fails to take into account the living organism's own contribution to its specific form. Goethe spoke of the particular individual plant as being a 'conversation' between the living organism and its environment. This metaphor draws our attention to the plant's *active* contribution to the form which it takes in specific conditions, emphasising the fact that the individual expression of the plant which we see is the outcome of the active response of the organism to the 'challenge' posed to it by the environment. This is stated very clearly by Steiner: 'We must conceive at a deeper level than the influences of external conditions something which does not passively allow itself to be determined by these conditions but actively determines itself under their influence.'[20] The living organism configures itself *actively*, instead of being conditioned passively, in response to the environment. The external conditions stimulate the plant but do not determine it. The plant responds actively out of its own 'potency to be otherwise' to produce the form of itself which the environment evokes. The Goethean scientist Craig Holdrege describes this as follows:

> Imagine that you are holding a groundsel seed in your hands before planting it. Depending on how, when and where you plant the seed, a limitless variety of forms can arise. All these potential forms are not, of course, stored in the seed. The concrete forms are emergent characteristics that arise out of a germinal state and develop in the interplay between the plant's plasticity and the environmental conditions. In particular surroundings the potential of the plant is evoked, but what appears is only one manifestation of the myriad ways in which this plant could develop.[21]

The specific form which an individual plant takes is neither determined by the environment nor predetermined by the organism itself.

As Holdrege indicates, we must avoid the trap of thinking in a 'finished product' manner, as if the potential forms were there already in the organism like peas in a pod. This is the kind of thinking which tries to 'get to the milk by way of the cheese', thereby eclipsing the dynamical quality of the organism be-ing itself differently according to the situation in which it is placed.

As well as the variety resulting from environmental factors, there is the much greater variety that can arise from the genetic variation taking place within the species. This is what interests the breeder. He or she is always on the look out for 'interesting' variations which can then be propagated – the process of artificial selection which Darwin took as his model for the idea of natural selection.[22] This is how the huge variety in any one species of plant arises. There are, for example, a thousand different varieties of Peony. Many of these are on display together on the same day at the Chelsea Flower Show in London. It is an astonishing variety to behold, and yet what we see before us extensively as many different plants is organically One plant which is intensively multiple – a 'multiplicity in unity' which is an expression of the dynamic unity of self-differencing. It is One plant be-ing itself differently and not just many different plants of a common kind. Of course, we usually see more in the 'downstream' mode of the latter than in the 'upstream' mode of the *living* plant. But if we can shift our thinking upstream, we can recognise that the diversity of peonies we see *is* the living unity of the Peony. How different it would be if we looked for unity among the peonies by trying to find what they all have in common. If what is living is always 'becoming other in order to remain itself', then we must learn to recognise diversity as the dynamic unity of life, so that we can see the unity concretely as being identical with the diversity of the phenomena. This is not what we would expect to find: that the unity is 'hidden' right in front of us as the diversity.

As we have considered the varieties of a single plant species as a whole, so we can go on to consider the different species within a genus – and then different genera within a family – also as a whole. It is an enlivening experience to see a particular family of plants in the light of the idea of the dynamic unity of self-differencing. We begin to see the different kinds of plant within the family intensively as One plant be-ing itself multiply, instead of just seeing different plants that have something in common. The extensive perspective, which remains on the outside of the phenomenon, tries to draw off the unity by abstracting what is the same, eliminating differences. The result is a lifeless 'unity in multiplicity' which is a dead end. We can recognise that the movement of thinking in this case is the opposite of that which sees the diversity of the phenomenon unfolding as the living unity of its coming-into-being. Anyone can learn to practise this living way of seeing for themselves. For example, we can become familiar with the different members of the *Rosaceae* family – the rose, cherry, apple, blackberry, strawberry, and so on – and begin to see them as One plant in the form of 'multiplicity in unity'. Learning to see in this way has the consequence that we begin to see each member of the family reflected in all the others, so that the rose is seen in the apple, as the strawberry is seen in the rose, for example, without there being any sense whatsoever that one kind of organism somehow changes physically into another. What we are seeing in this way is the metamorphosis of One plant into different modes of itself, and not the external change of one plant into another one. How different the *experience* of this is from that of looking for what these different plants have in common, i.e. from seeing the *Rosaceae* in the mode of the static unity of self-sameness. This latter way of seeing leads only to an abstract generalisation – something like an 'average plant' of the *Rosaceae* family – which functions as an organisational schema, or 'blueprint', for all the plants of the family. Although such a concept has been used in biology at times – and has often been mistakenly identified with Goethe's approach – it is nevertheless so rigid and static that it completely lacks the flexible and dynamic quality which is characteristic of life. It is in fact no more than a lifeless counterfeit of living being, the very opposite of Goethe's living perception of nature for which he says 'we must make ourselves as mobile and flexible as nature herself' (see Figure 9).

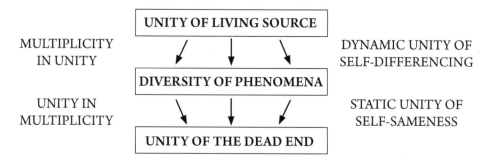

Figure 9. From the living source to the dead end.

The Archetypal Movement

The dynamic unity which is the plant is to be found at every level: the organs of the plant, the varieties of a single species, the species within a family, and ultimately the One plant which is the whole plant kingdom – what Goethe called the *Urpflanze*. It is very important that we try to think dynamically when we are considering this notion, going 'upstream' into the coming-into-being, and do not fall back into the habit of static thinking which begins 'downstream' with the organisms in their finished state. Much of the misunderstanding which there has been about Goethe's *Urpflanze* is a result of failing to do this. Time and again we read that Goethe searched for the general plan common to all plants, looking for uniformity and commonality in the multiplicity of living organisms, trying to reduce the diversity of nature to unity, and so on. But this just gets what he was doing back to front. It is only by thinking in a dynamical way that we can come to appreciate Goethe's notion of the *Urpflanze,* and the way that it differs significantly from some of the misinterpretations to which it has frequently been subjected.

The name *Urpflanze* is usually translated into English as either 'primordial plant' or 'archetypal plant'. Both of these *could* be seen in the Goethean way – i.e. as referring to the intrinsically dynamic unity of coming-into-being which is the One plant be-ing itself differently as 'multiplicity in unity' – but almost invariably they are not seen in this way. The term 'primordial plant' is usually taken to refer to some supposed primitive ancestral plant from which all other plants have developed procreatively over time. The use of the adjective 'archetypal' immediately suggests an association with the philosophy of Platonism. In the standard established interpretation, this is a two-world philosophy and hence strongly dualistic. According to this interpretation, which has been held widely, Plato conceived the fundamental ontology of the world as being on two different levels. There is the level of what we see around us, the world of changing appearances that we experience through the senses, and the ontologically superior level of the world of Ideas or Forms which are eternally one and the self-same, and constitute the original 'templates', or 'models', according to which the multifarious appearances of the world of the senses are formed, and of which they are only imperfect copies.[23] In the standard interpretation, these two different ontological levels are conceived as being not only distinct but also as being *separated* from one another – hence the two-world dualism. For all the *many* different instances of some particular kind of thing in the sensible world, there is *one* Form or Idea in the intelligible world which functions as the ideal archetype according to which all the particular instances are formed – this is referred to as 'the one over many', where a *separation* is implied between the one and the many. It seems clear that this is a very external, spatial way of thinking: another world is imagined which seems to be 'outside' the familiar world, and yet this second world is imagined in the image of the familiar world, which it therefore seems to duplicate in an idealised way. The difficulties to which this leads are well known, and are so intractable that we cannot help but wonder why Plato ever thought this way in the first place. The answer may well be that he didn't. What we think of as Plato's philosophy may really be a misinterpretation of Plato. It is possible that Plato himself did not subscribe to the two-world theory implied by this common (mis)interpretation of the theory of Ideas or Forms – which Gadamer, who affirms that 'Plato was no Platonist', refers to as pseudo-Platonism, or vulgar Platonism. However, Plato himself may have unwittingly encouraged this misunderstanding by the way that he presented the Ideas, and their relation to the world of the senses, and later he went to considerable lengths to correct it.[24]

We can see from all that has been said about Goethe's dynamic way of thinking, that what he means by the *Urpflanze* has nothing whatsoever to do with the notion of the archetype in the standard interpretation of Plato. For this reason, translating it as 'archetypal plant' can be very misleading, because it invites us to associate it with a notion which is completely out of tune with his whole way of thinking.[25] What Goethe means by the *Urpflanze* is the dynamic unity of the coming-into-being of all plants as the self-differencing of One plant, which is therefore intensively multiple but appears to us extensively as all the many different plants. What this means is that each plant *is* the *Urpflanze* be-ing one possible mode of itself – the number of possibilities is indeterminate. Hence, paradoxically, it is everywhere visible and nowhere visible – although once we begin to think dynamically, this is no paradox at all. Instead of being *separate* from the many particular plants that we see, i.e. as 'the one over many', Goethe's *Urpflanze* is One which comes into concrete manifestation simultaneously with the many – with which it is identical because the many are now the self-differences of One. This is very different indeed from the two-world theory which *separates* the One from the many. There is no such dualism in Goethe's thinking, for which, in his own words: 'The universal and the particular coincide: the particular is the universal, appearing under different conditions'.

Precisely the same can be said about the *Urorgan,* Goethe's 'diversely metamorphosed organ' coming-into-being as the self-different organs of the plant. This is often translated as 'archetypal organ', bringing with it the kind of misunderstanding just described. However, it is possible to use the term 'archetypal' – as opposed to just leaving *Urorgan* and *Urpflanze* untranslated – if we are careful to think dynamically. There is no archetypal *entity* – whether an organ or a plant – but there is an archetypal *movement*. In fact this is what we have been describing all along: the archetypal movement is the *intensive* movement of self-differencing. So we are describing the plant in the archetypal mode whenever we see it as the dynamic unity of self-differencing – whether it is the single plant organism, the variety of plants, or the whole plant kingdom. The key is that the 'archetype' is not an entity but 'a movement in which it is one and yet becomes different at the same time'.[26] If we were to perceive the diversity of plants in the archetypal manner, we would have a purely dynamic experience of seeing different plant forms appearing one after the other, as if they unfolded out of one other. Goethe describes such an experience:

> When I closed my eyes and lowered my head, I could
> imagine a flower in the centre of my visual sense. Its original
> form never stayed for a moment; it unfolded, and from
> within it new flowers continuously developed with coloured
> petals or green leaves.[27]

We must read this intensively, as One plant coming-into-being self-differently, and not extensively as one plant after another.[28]

Unfortunately, Goethe's dynamic archetype is all to often mistaken for the abstract universal of a static generalisation. We can see this confusion in the way that comparative morphology developed in Britain in the Victorian period, especially in the work of the famous anatomist Richard Owen, director of the Kensington Natural History Museum in London. In his major work on the vertebrates, *On the Archetype and Homologies of the Vertebrate Skeleton* (1848), Owen set out an idealised picture of the simplest vertebrate form. But this is not the unity of the living source of all potential variations. On the contrary, it is the most basic pattern common to all vertebrates, the least common denominator shared by all members of the vertebrate class. Such an abstract generalisation is evidently at the opposite pole to what Goethe had in mind. This reduction to the minimal commonality from which all the specialised organs required by actual living organisms have been excluded, is really the unity of the dead end. Owen and others believed that in this way it would be possible to discover the most general ground plan (the 'blueprint') of the vertebrates – which is what he called the 'archetype' of the vertebrates. There is, of course, nothing wrong with such an abstract generalisation; what matters is the use to which it is put. In fact it can be very helpful in the process of discovery. For example, it was in this way that Owen came to the concept of homology, whereby different organs – for instance, the bones in the human hand, the wing of the bat, and the paddle of the porpoise – can all be recognised as instances of the same basic anatomical structure fulfilling different functions.[29] There were many anatomists who just saw this kind of abstract unity of organisation common to different organisms as no more than a geometrical abstraction – it was even sometimes compared to a mathematical axiom.[30] But Owen was at the forefront of those who wanted to see more in it than this. He wanted the unity of the *abstract* universal to be a *transcendent* unity, as if it pre-existed as the ground plan of the vertebrate class at a 'higher' ontological level than the actual organisms themselves. We can begin to recognise the two-world theory of Platonism here – which, as already mentioned, is really pseudo-Platonism.[31] This is very different from Goethe's idea of the science of morphology, which can only be understood in a thoroughly dynamical way and not in terms of static 'Platonic' archetypes. Because his way of thinking is intrinsically dynamic, it is possible to describe the dynamics of being without introducing a false dualism which separates being from appearance. But to do this we have to go upstream into the coming-into-being, instead of beginning downstream with the finished organisms.

We can of course begin at the end by abstracting the unity of what the finished organisms have in common. As we have seen, there is nothing wrong in doing so. But we also need to keep our attention on the movement of thinking, otherwise we will make a fundamental mistake which has far-reaching consequences – which is the mistake that Owen and others made when they turned the abstract universal into a transcendent unity. Having formed this abstract unity, which can only come at the end, we then project it back into the origin, and imagine that it is there in the phenomena, or 'behind' the phenomena, all the while. In other words, we assume that what is in fact a downstream abstraction is ontologically fundamental. In which case we now have to try to understand how difference could emerge from an abstract unity from which all difference has been excluded. It is impossible. Since all difference has been excluded from this unity – in favour of what is common – then none can emerge from it. It is an ontological cul-de-sac. Yet this mistake has been made time and again. What this misses is the dynamic unity of the living source, the unity of coming-into-being, for which it substitutes the static unity of the dead end. The result is that we try to 'reach the milk by way of the cheese,' and so get everything the wrong way round. Goethe's understanding of the dynamics of being, on the other hand, goes upstream towards the unity of the living source, so that the movement of his thinking follows the coming-into-being of the phenomena and ends where we usually begin. Consequently in Goethe's dynamical thinking of 'the one and the many' there is no separation of the One from the many, and the two-world dualism of pseudo-Platonism simply doesn't arise.

Modes of Unity and the Bimodal Brain

It is at first surprising that views have been attributed to Goethe which are in fact the very opposite of his own vision of the dynamic unity of nature. We can understand how the idea of the minimal commonality arises from the movement of thinking which begins downstream with the finished organs/organisms, and then proceeds further downstream to abstract what they have in common with one another, whereas what is required is to reverse this direction of thinking and go back upstream to follow the coming-into being of the organs/organisms. Even those who do emphasise that Goethe's approach should not be confused with that of other anatomists – Richard Owen, for example, and indeed the overall approach in British anatomy in the Victorian period – nevertheless often fail to appreciate sufficiently the *intrinsically* dynamic quality of Goethe's way of thinking. For example, Robert Richards, in his magisterial work, *The Romantic Conception of Life,* says of Goethe that:

His efforts would also differ from those later anatomists, like Richard Owen, who would pursue a general archetypal pattern but one that illustrated the least common denominator of the vertebrate class.

before going on to say:

By contrast, Goethe conceived the archetype as an *inclusive form,* a pattern that would contain all of the parts really exhibited by the range of different vertebrate species.[32]

Such an 'inclusive form' is certainly an improvement on the 'minimal form' of the lowest common denominator. It is a step in the right direction, but doesn't go far enough. Thinking of an inclusive form leads us to imagine it 'as including all of its possibilities', as if it already 'contained' all its potential variations. We can recognise that this is really thinking in a 'finished product' manner, as if the variations were already there, like peas in a pod waiting to come out – we have already discussed the snare of this kind of thinking earlier in this chapter. It is only too easy to fall into the trap of thinking in this way, whereas to reach Goethe's dynamic way of thinking requires us to take a further step. Richards himself almost takes this step when he says:

With the mental eye, we would see form as dynamic, as containing its infinite variety of transformations within a unity.[33]

Yes the form is dynamic, but this is not described adequately by saying that it contains its variety within a unity – this is already 'too late'. But then it is very difficult to indicate the dynamical quality of Goethe's organic thinking, and only too easy to describe it instead in a downstream way.[34]

The failure to recognise the dynamical character of Goethe's way of thinking is certainly at least in part a consequence of the dynamics of thinking itself. The intrinsic direction of 'coming-into-being' is towards what thus comes into being, and so it is inevitable that consciousness is fixed on the end result. In latching onto the outcome it thereby overlooks the dynamics. In this way the dynamics of thinking promotes its own eclipse. This is well-known in phenomenology, where in attempting to re-enter experience as lived we become aware of the tendency to use the results of experience to account for that experience itself.

But there is more to it than this. We saw in the previous chapter that we tend to rely mostly on the verbal-intellectual mind, which functions in terms of abstract generalities, whereas Goethe redirects attention into the concrete particularity of sense experience. He does this by active looking and the practice of exact sensorial imagination. It is through this activation of the sensory that we come to encounter the phenomenon in a dynamic way, so that we begin to experience the phenomenon as coming-into-being, whereas relying on the intellectual mind only brings us into contact with the phenomenon as it has already become. The verbal-intellectual mode of apprehension looks for what things have in common, the respect in which they don't differ at all, which leads to the mode of unity of the abstract universal from which difference is excluded. This is the *form* that our thinking of 'unity' usually takes because we are relying on the intellectual mind. But if this is the mode in which we approach Goethe, then clearly we will *only* be able to understand the 'unity' of nature in this way, which completely misses the mode of unity to which Goethe is drawing our attention.

In the final part of the previous chapter, it was suggested that the difference between the verbal-intellectual and the sensuous-intuitive modes of experience can be correlated with the left and right hemispheres of the brain. To begin with, it was thought that the difference between the hemispheres could be understood functionally in terms of 'what' each hemisphere did – the left being analytical, whereas the right is holistic. So, on the basis of 'what' each hemisphere does, the idea of 'unity and multiplicity', or 'the one and the many', would be partitioned between the hemispheres, with 'unity' being mediated through the right hemisphere and 'multiplicity' through the left. But it turns out this is entirely the wrong way to think about it. What seems to be the case is that 'unity and multiplicity' is mediated through *both* hemispheres, not split into 'unity' through the right hemisphere and 'multiplicity' through the left. So instead of asking 'what' is mediated through each hemisphere, the question we should ask is 'how' is 'unity and multiplicity' mediated through the left hemisphere and 'how' is it mediated through the right? By 'how' is meant 'the manner in which' and not 'the means by which'.[35] So the question is, what is the manner in which 'unity and multiplicity' is mediated through the left hemisphere, and what is the manner in which it is mediated through the right?

How is 'unity and multiplicity' mediated through the left hemisphere? This is the hemisphere of abstraction in which things are taken out of context and so appear as separate entities. At the same time it is the hemisphere which produces generalisations, abstract types, and classifications. So the left hemisphere excludes difference to see unity in terms of what the multiplicity has in common. This is the manner in which 'unity and multiplicity' is mediated through the left hemisphere, the *aspect* of 'unity and multiplicity' which appears. Where the left hemisphere 'sees' things abstracted from their context and separated into parts, the right hemisphere 'sees' the whole before it gets separated into parts. It also sees things in their context:

> For the same reason that the right hemisphere sees things
> as a whole, before they have been digested into parts, it also
> sees each thing in its context, as standing in a qualifying
> relationship with all that surrounds it, rather than taking it as
> a single isolated entity. Its awareness of the world is anything
> but abstract.[36]

The right hemisphere sees things concretely, in a way that can only seem paradoxical to the left hemisphere. Because it sees concretely instead of abstractly, it is 'more concerned with the uniqueness and individuality of each existing thing or being'. Yet at the same time it does not see things separately but 'sees them as belonging in a contextual whole, from which they are not divided'.[37] It is this awareness of individual characteristics which simultaneously belong to a whole that seems paradoxical. Difference stands out now, instead of receding into the background as it does in the sameness of left-hemisphere generalisation. Yet this difference is within wholeness. So if we ask how 'unity and multiplicity' is mediated through the right hemisphere, we can see now that it as 'multiplicity in unity', in which multiplicity appears within unity (i.e. without fragmenting it), instead of unity being abstracted from multiplicity, as it is in the left-hemisphere form of 'unity in multiplicity'.

There is a sense of reversal in going from one mode of unity to the other, as if each mode of unity is like the other turned inside out.

Looking back to the example given towards the end of Chapter 2 – of my own experience of seeing the difference between 'unity in diversity' and 'diversity in unity' in the context of a display in a museum – we can now realise that going from one way of seeing to the other is an exercise in going from the left to the right hemisphere of the brain. If we practise going backwards and forwards between these two ways of seeing, we can experience the world appearing in different aspects. Whenever we make the effort to go 'upstream' in our thinking into the coming-into-being, we are focusing through the right hemisphere of the brain. Whereas when we begin with the end result, and go further 'downstream' into abstract generalisation, we are focusing through the left hemisphere. So the difference between these two modes of unity reflects the difference between the two hemispheres of the brain.

4. The Philosophy of Unfinished Meaning

The dynamic approach tries to 'catch things in the act', in the lived experience, instead of 'after the fact' in the 'replay' of representation – which we then mistake for the experience itself. In this chapter we are going to consider the *phenomenon* of understanding the meaning of a written text. This will bring us to the hermeneutic philosophy of unfinished meaning, where we will find that the meaning of a work can be one and yet many at the same time without this leading to the dichotomy of objectivism and relativism. By shifting our attention 'upstream' from the meaning that is understood into the coming-into-being of meaning in understanding, we discover the same form of the dynamic unity of self-differencing that we have seen in Goethe's account of 'the one and the many' in the living plant.

The Common Sense View

The question we are concerned with is what happens when we understand the meaning of a written text? Putting it in a more Kantian manner, we could ask what are the conditions for the possibility of understanding a text? But whichever way we put the question, the biggest barrier to giving an answer which is adequate – i.e. one which does justice to the experience instead of the presuppositions through which we subsequently filter it – is that we usually approach it in the framework of the subject- object model of experience, which *begins* by *separating* subject and object. The most widespread view is one which locates the meaning of a work in the author's original intention – i.e. what the author 'had in mind' when he or she wrote it. It seems to be just common sense that the author had something he wanted to say, which he expressed in the written text, and consequently that we understand the text if we can retrieve from it what the author meant. If we are trying to understand *The Critique of Pure Reason,* for example, we want to know what Kant meant, what he was trying to say. Thus it seems that, if we identify the meaning of the work with what it meant for the author, understanding the work becomes a matter of somehow recovering the author's meaning from it, of reproducing the original intention of which the work is the expression.

There are a number of concomitant aspects of this 'common sense' realism which need to be brought out explicitly. In the first place it supposes that there is an 'objective' entity – the author's meaning – which as such exists independently of the reader who wishes to understand it. Furthermore, this objective meaning is stable and determinate:

> When ... I say that a verbal meaning is determinate I mean
> that it is an entity which is self-identical. Furthermore, I also
> mean that it is an entity which always remains the same from
> one moment to the next – that it is changeless.[1]

If the meaning of the text is the author's meaning, and as such is therefore invariant (self-identical and unchanging), then the meaning can be understood only by being *reproduced* in the reader's experience. So whereas the meaning of the text is located in the author, the understanding of this meaning is located in the reader. The meaning is treated as being an object ('in the author's mind') which is understood when it is reproduced or copied in the subject ('in the reader's mind'). Evidently what we have here – which follows more or less of necessity from the assumption that the meaning is effectively an invariant object – is the classical epistemological position of the subject-object separation and the concomitant question of how this can be overcome. Transposing into the more familiar, traditional epistemological question of how we know the world – how do we know there is a glass on the table? – we can recognise immediately that common sense realism 'epistemologises' hermeneutics, so that the all too familiar problem of scepticism which arises with the representational theory of knowledge is now transferred also to the question of how we understand the meaning of a text. How can we know that what is in the reader's mind is in fact what the author had in mind? In other words, how can we know that it *is* a reproduction of the author's meaning? How can the subject-object dichotomy in which we have unwittingly trapped ourselves be overcome? The answer – or at least part of it – is that it cannot be overcome once it has taken hold of us because then it is too late. But before going into this further, we should note the logic of the concepts here in the way they are internally related:

> Reproducibility is a quality of verbal meaning that makes interpretation possible: if meaning were not reproducible, it could not be actualised by someone else and therefore could not be understood or interpreted. Determinacy, on the other hand, is a quality of meaning required in order that there *be* something to reproduce.[2]

In other words, if *reproducibility* is the necessary condition for understanding to be possible, then it follows that meaning *must* be invariant. Concomitantly, if meaning is invariant, then it follows that understanding can only be reproductive. The one necessarily entails the other.

If we identify the meaning with 'what the author had in mind', then clearly we have a separation between meaning and understanding:

objective meaning	*subjective understanding*
in the	*in the*
author's mind	*reader's mind*

Because we cannot have direct access to the original meaning, we have the problem, which is unavoidable within the subject-object dichotomy, that we cannot be certain that our understanding is correct – i.e. that it corresponds with what the author had in mind (notice that here truth is reduced to correctness). We cannot leap over the reproduction in the reader's mind to check that it corresponds with what the author had in mind – i.e. that what is in the reader's mind really is a *reproduction* of the original meaning (this is the hermeneutic version of the egocentric predicament in epistemology). Our only way of access to the meaning is through the text.

Furthermore, there is the undeniable fact that different readers can, and often do, understand one and the same text differently. So if the meaning of the text is what the author had in mind – what the author intended – then clearly only one (or none) of these readers understands the text correctly. But how do we know which one (if any) this is, and what status can be given to those readings of the text which seem coherent and plausible but which are judged not to be correct, i.e. not what the author intended? Are these simply to be rejected, or do they have some secondary value? This question becomes especially important when we consider how the way that a work is understood changes over the course of time. It seems that the meaning of the work changes with time, but this is clearly impossible if the meaning is what the author originally intended.

One way of attempting to overcome these difficulties is to introduce a distinction between 'meaning' and 'significance'. This is an attempt to distinguish the meaning which the text has in itself, as a fully determinate entity which is self-identical at all times, from its meaning for us which will change with time and circumstances. Whereas the significance of the text will vary according to historical and cultural context, this cannot affect the meaning of the text which is effectively timeless:

> The historicity of all interpretations is an undoubted fact,
> because the historical givens with which an interpreter
> must reckon – the language and the concerns of his
> audience – vary from age to age. However, this by no means
> implies that the meaning of the text varies from age to age,
> or that anybody who has done whatever is required to
> understand that meaning, understands a different meaning
> from his predecessors of an earlier age ... the historicity
> of interpretation is quite distinct from the timelessness of
> understanding.[3]

This distinction between meaning and significance has the effect of 'saving the appearances' of the separation of subject from object. By separating 'objective' meaning from 'subjective' significance it seems on the face of it to accommodate the problem of differences in interpretation without sacrificing the invariant meaning of the text. But as soon as we look into it more closely, we discover that this distinction is not as helpful as it might seem at first.

It has the undesirable result of effectively placing the meaning of the text beyond the possibility of experience, attributing to it a state of timelessness in which it is protected forever from change – which therefore, as well as excluding change which is accidental to the meaning, also excludes the kind of change which is the further realisation of the possibility of the meaning itself (we shall consider this in detail below). Understanding is thereby effectively restricted to the changing significance which the work may have for us in different times and circumstances. So, far from being helpful, the distinction between meaning and significance condemns us to an extreme form of dualism, which removes the meaning from the changing world of human experience and puts it into a timeless state of permanent self-identity in which it is always the same under all circumstances and for everyone – with the result that 'one would have to possess the timeless mind of a Thomistic angel' to understand it.[4] The significance which the work has for us, by contrast, is no more than the changing shadow of this meaning cast in the light of time and circumstances. This is surely a very strange way to save the objectivity of meaning from the relativism of subjectivity: to make it inaccessible! It seems to deny the very possibility of understanding.[5] So it turns out that, what seemed like a way of accommodating the consequences of the subject-object dichotomy, has the opposite effect of only digging us in deeper. The distinction between meaning and significance only serves to exacerbate the difficulty, because it reinforces the very dichotomy which leads to the problem of objectivism and relativism in the first place.

The Dynamic Approach

However, there is an alternative approach. In all that we have considered so far, we have taken meaning as *finished* meaning – what the author had in mind. In doing so we have begun with what is already a finished product, which somehow has to be recovered or reproduced in the mind of the reader for it to be understood. We can now easily recognise that this approach tries to begin 'downstream' with finished meaning, which is conceived as an end product that exists in itself entirely independently of whether or not it is understood by someone other than the author. This is why we find ourselves unavoidably in the grip of the subject-object separation. The alternative here is to reverse the direction of attention, which is drawn naturally to the end product, by stepping back 'upstream' into the *event* of understanding. The effect of this is that meaning emerges in the *happening* of understanding, instead of being present already as a finished object *before* it is understood by the reader. If we can shift our attention upstream in this way, we find ourselves prior to the separation of meaning from understanding and hence before the separation into subject and object. This brings us to the phenomenology of the event of understanding.

One of the factors which makes phenomenology seem obscure is the very name itself – which one contemporary phenomenologist has described as 'misleading and clumsy'.[6] Faced with the term 'phenomenology', instead of being led into the *experience* to which it refers, we are misled into a search for the meaning of the term itself, and before we know it we are ensnared in all the intellectual paraphernalia which philosophy seems to produce so easily. What it really refers to is a movement of thinking in which the position of attention is shifted from *what* occurs (downstream) into the *occurring* of what occurs (upstream). In particular, it is concerned with the happening of appearing – with appear*ance* (read verbally) – so that phenomenology is concerned with what appears in its appearing. As we saw in the first chapter: 'when one speaks of what "appears", one refers not only to a *thing* but to a *happening*: the appearing itself', so that 'a phenomenon is not only something which appears, but something which appears *as appearing*'.[7] Clearly there cannot be any separation between the happening of appearing and what appears – i.e. there could not be 'appearing' without 'something' appearing. But our attention is usually drawn to *what* appears to such an extent that we miss the happening of appearing. In fact, although it clearly makes no sense to try to think of appear*ance* without something thus appearing, we almost invariably do think of *what* appears without noticing its appear*ance*.

As we explore the shift in attention which this requires, to catch what appears in its appearing, we find ourselves in a position where familiar patterns of thought that we take for granted no longer apply. When we focus in the usual way on what appears, it seems just natural to say 'it appears'. But when our attention shifts upstream into what appears in its appearing, then it becomes awkward to say 'it appears' because the very form of this leads us to think of an 'it' which 'appears'. This encourages us to think of 'it' as being there already, and then appearing. But this gets it back to front, by imagining 'it' as if it had already appeared before it 'appears'! We would do better to say 'appears it'. This may be bad grammar, but it is better philosophically because now 'it' emerges *for the first time* in its appearing, and so this avoids the mistake of separating 'it' from 'appearing' as if appearing is something that happens to 'it' subsequently. This further implies that appearing is contingent to 'it', in the sense of being something that sometimes happens to it but need not necessarily do so.[8] Directing our attention into the movement of thinking in this way, enables us to see clearly the difference between 'it appears' and 'appears it', and to recognise that the self-contradictory character of the former encourages us to get everything the wrong way round.

But this is by no means all that we discover when we shift the position of attention back 'upstream' in our experience. Surprising as it may seem at first, we discover that there is no separation between appearing and seeing. To begin with, even without noticing it, we think of appearing and seeing separately, as if appearing happens first followed by seeing. But this is the 'downstream' way of thinking, according to which we try to think of something as if it first appeared and then is seen – i.e. as if seeing is subsequent to appearing. But this would mean there could be appearing without seeing – and therefore that seeing is not necessary for appearing. Yet the very notion of appearing necessarily entails seeing – try to imagine appearing, i.e. the happening of appear*ance*, without seeing. If something appears, it must be seen, otherwise it could not 'appear'. Not only could something not appear without being seen, but also for that very reason it could not appear first and then be seen. When we try to catch it 'in the act', instead of after it has happened, we find there can be no separation because appearing *is* seeing. There are not two events, an event of appearing and an event of seeing, but a single event which could be described equally well as appearing or as seeing. We shall call this the unitary event of {appearing/seeing}(see Figure 10).

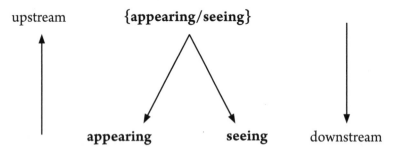

Figure 10. The event of appearing.

It is only when we begin at the end, with what has appeared instead of with its appear*ance*, that appearing and seeing seem to be separated.

They are distinguished, but the distinction is intensive, whereas their separation would be extensive.

Aristotle and the Unitary Event

We can see now that there is a *single actualisation* (upstream) where we would usually think in terms of two different events (downstream). In fact, this 'single actualisation' is at the heart of Aristotle's philosophy – beginning with his description of change in the *Physics,* and continuing through the account he gives of perception and understanding in *De Anima (On the Soul).*[9] What is fundamental for Aristotle is the notion of change *as such*, which he says is 'the actualizing of potential being as such'. We must be careful to read this dynamically:

> One might well think that the actuali*zing* of a potentiality would be the finished product: for example, the actualization of what is potentially a man would be a man. But to speak of the actuali*zing* of a potentiality *as a potentiality* is to isolate the process by which the potentiality turns into an actuality.[10]

So, in the activity of building for example: 'When the build*able* (*to oikodometon*) is actualized, he says, it is *being built* (*oikodomeitai*) and this is *the process of building* (*oikodomesis*)'.[11] So on this account, the actual house that is built is *not* the actualization of the potential *as such* – although this is how we would think of it today. Once the house is built, there is no longer an actualising of the potentiality *as such* because there is now no potentiality to be actualised. Aristotle's thinking is always dynamic.

Although this description of change may seem somewhat 'abstract' to us at first, it is in fact the very opposite. It is a concrete description of how Aristotle experienced change in both the natural world and the world of human experience. Today we think of an 'event' as something happening at a particular place and time, i.e. we see it in the context of a spatio-temporal matrix even without our realising it. But, as Lear points out, Aristotle didn't have a watch, which means that he couldn't specify an event in the way that we now (after the seventeenth century) take for granted. It is really our way of thinking of an event which is abstract, whereas Aristotle's way of characterising an event in terms of the actuali*sing* of a potentiality *as such* is much closer to concrete experience. It is a consequence of this description of change that there is a *single* activity or event which is simultaneously cause and effect:

> For example, a pile of bricks may be a house potentially and a builder may be able to build the house. The actualization of these potentialities is the building of the house. Indeed, Aristotle says that change can be understood as the actualizing of the potential agent and the patient. Thus we can think of a change in terms of a builder actualizing his potential by becoming a builder building and the bricks having their potential actualized by becoming a house being built. However, the actualizing of these two potentialities is not two separate events. The actualizing of the agent and the actualizing of the patient are, for Aristotle, one and the same event.[12]

So what we have is a unitary event of {builder building/house being built} in which the 'builder building' and the 'house being built' both refer to one and the same event but in different ways – we could perhaps think of it in terms of the duck/rabbit gestalt. The consequence of this dynamic description of change, which leads to the single actualisation, is that cause and effect cannot be separated into two events – as we have become accustomed to doing since the rise of modern science in the seventeenth century – because 'The event which 'the builder building' refers to is every bit as much the effect as it is the cause'.[13]

Aristotle then goes on to consider the question of *where* the actualising of the agency is located. The answer he gives is that the actualising of the agency is located in the patient. We can easily see this in the activity of building. Here the activity which is the actualising of the agent, i.e. the builder building, must be occurring in the patient, i.e. the building being built. Where else would the activity of the builder building be located except in the materials which are becoming a house? But Aristotle goes on to consider another example, where the consequence of the single actualisation description of experience is more surprising. In *Physics* III.3, he asks us to consider the case of a teacher teaching and a student learning.[14] Where we would see this as *two* separate but related activities, Aristotle sees it as a single activity which can be described either as 'the teacher teaching' or 'the learner learning'. But if the actualising of the agency is located in the patient, this means that the teacher's teaching is occurring in the student!

This seems strange to us, because we imagine two activities, and so we believe we can imagine the teacher teaching without the student learning – in fact, those of us who have been involved with education might well be tempted at times to think this is the norm! But in this case Aristotle would say that the teacher isn't *teaching* – and if he is *teaching*, where else could this occur but in the student, and *this* is what we call the learner learning. In fact, in the single actualisation description, the teacher teaching *is* the learner learning. It is the unitary event of {teaching/learning}. So if we now go back to the unitary event of {appearing/seeing} and ask *where* is the appearing, the answer is that it is in the seeing. In fact 'appearing' and 'seeing' are one and the same event: the appearing *is* the seeing – this is why we call it the *unitary* event of {appearing/seeing}.

The Event of Understanding

With this preparation, we can now return to the dynamics of the event of understanding. The alternative approach is clearly to shift attention away from meaning as an object into meaning as an event – i.e. into the happening of meaning. The term 'meaning' in English is a participle, so that it functions as either a noun or a verb. In the first mode it gives us the sense of *a* meaning as an entity. But in the verbal mode, it gives us the sense of meaning as intrinsically dynamic, as mean*ing* instead of *a* meaning. In this case we can begin to catch 'meaning' in the active sense of the happening of meaning, i.e. in the sense that meaning *means* (read verbally). It would be useful to distinguish 'active meaning' or 'working meaning', from 'object meaning' or 'finished meaning'. It is only the latter which we are thinking of when we consider meaning as if it is present as a determinate entity. The expression 'working meaning' is particularly helpful because it goes together with the notion of the text as a 'work'. Although this is usually taken to refer to the text in the sense of being the author's work, and hence to something that has been done, it can also be used in a more immediate way to refer to the active working of the text which we encounter in the event of understanding. So that in this case 'the meaning of the work' should be read actively as the work meaning in the event of understanding.

The event of meaning is the appearance of meaning. Following what has been said about the unitary event of {appearing/seeing}, we can now recognise that what we are concerned with in this case is really a specific instance of this, namely the unitary event of {meaning/ understanding}. When we shift attention upstream we do not find a meaning which is already there, as if it had happened already, but the happening of meaning – we catch it 'in the act' of meaning. As we cannot have appearing without seeing, so also we cannot have meaning without understanding. The meaning does not appear first, and then we understand it. Understanding is not a response to a meaning which is there already; it is the appearance of meaning. So we can say that the appearing of meaning is the happening of understanding. When we try to catch it 'in the act' we find there is no separation between meaning and understanding. There are not two events, first an event of meaning and then an event of understanding, but a single event which could be described equally well either as meaning or as understanding. This is the unitary event of {meaning/understanding}, which is prior to the subject-object separation. It is when we slip downstream from this event, as attention shifts to focus on *what* is understood, that this separation happens (see Figure 11).

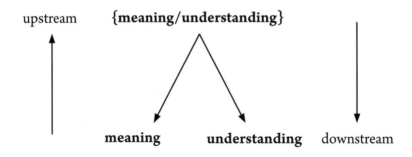

Figure 11. The event of meaning.

Meaning and understanding seem to be separate only when we begin with meaning that is finished, instead of with the lived experience in which meaning comes into being as the happening of understanding. The unitary event separates into meaning *and* understanding when our attention falls from the happening of understanding, which is the appear*ance* of meaning, to *what* is understood. This shift of attention is inevitable, but the consequence of beginning with *what* is understood is that we begin with what has already appeared, with the result that we *separate* meaning and understanding and as a consequence find ourselves in the subject-object dichotomy.

The unitary event of {meaning/understanding} is an encounter with meaning which is *active* because it is happening *now*, and not the reproduction of an original meaning which is finished and comes down to us as such from the past. As Gadamer says, 'understanding as reconstructing the original would be no more than handing on a dead meaning'. Understanding is therefore not 'the reproduction of the original production', but the event in which, what otherwise would be only 'the dead trace of meaning', is 'transformed back into living meaning'.[15] A text is not a memorial to the past, but a doorway through which the past can come to life for us now. This coming to life again in the present, instead of just being handed down to us from the past, is quite extraordinary – Gadamer calls it a miracle – and at some time we will all have experienced it as such, in the instant before it becomes covered over with layers of familiarity and begins to seem ordinary again. If the meaning, of which at first we have only a dead trace, comes to life now and becomes *present* meaning, then clearly it is contemporaneous with us. In such an event there can be no separation between meaning and understanding, because the meaning comes to life *as* the happening of understanding. If 'understanding must be conceived as part of the event in which meaning occurs', then meaning is not self-contained, simply there to be discovered, but 'is always *coming into being* through the "happening" of understanding'.[16] If meaning is actualised only when the text is understood – the unitary event of {meaning/understanding} – then clearly there is no separation here into subject and object. In which case meaning is not present as an object to a subject, but comes-to-presence in the event of understanding. It is *after* the meaning has been understood that it is represented in the second-order description as if it were an object, and is thereby 'read back' into the event of understanding as if it had been there all along.

The Hermeneutic Reversal

Although the reader will be able to draw on his or her own experience of meaning coming into being in the event of understanding, it will nevertheless be useful to give a specific example of the experience of meaning coming to life. As Gadamer says, 'it is universally true of texts that only in the process of understanding them is the dead trace of meaning transformed back into living meaning'.[17] The following instance of this is taken from Jason Elliot's account of his travels in

Afghanistan, *An Unexpected Light*.[18] Although this example does not entail a written text in the conventional sense, Gadamer shows in *Truth and Method* that, in a way which is similar to the meaning of a written text, the meaning of any work of art is actualised only when it is understood. The example given below, although not simply a text in the conventional sense, is nevertheless a 'literary work of art'.

The author describes his visit to the shrine of a Sufi Saint, one which he had particularly wanted to see. The conditions weren't good; the light was fading as they reached it, and the question of their safety when it got dark began to become pressing. Nevertheless, he and his companion entered the shrine, and when they did so he found that the place itself seemed to take over his experience in a subtle way. He found one part of the elaborate decoration behind the tomb particularly striking:

> The section, about twenty foot square, was made of
> tessellating panels in the form of squares and rectangles,
> the centres of which alternated in dark and light tile against
> a middle shade in each different panel, like positive and
> negative photographic images. What appeared from a
> distance to be the shading within these shapes was in fact
> a mosaic of angularly stylized Arabic characters, with each
> character itself composed of tinier tiles. The contrast in
> colour lent the panels a three-dimensional effect. I found
> my eye drawn outwards, and realized that the interlocking
> borders of each panel were formed by the extended lines of
> swirling black lettering that enclosed the entire design. The
> angular and the cursive had thus become inseparable, without
> having sacrificed their own strong identities which, alone,
> would have overpowered the overall sense of the design. That
> the mosaics themselves depicted verses from the Qur'an
> added another dimension of significance to the whole, and
> as if to mirror the paradox suggested in the shapes, allowed a
> changing meaning, in words this time, to express itself within
> an unchanging form.
>   Above this section ran a wide band of bright, densely
> overlapping calligraphy in white against royal blue. And
> above this came another panel as extraordinary as the first.
> Here, interlocking squares and triangles – each, again,
> displaying an individual design – had been ingeniously
> skewed to give a strong impression of three dimensions.

The pattern of tiles here must have been conceived of in three planes before it was expressed in one. The resulting shapes seemed to tease the eye like a puzzle, and the squares looked like the sides of unfolding cubes, which shifted in and out of one plane and into another before I could quite decide to which they really belonged. I felt stunned; they seemed to detach themselves from the wall and began to open and close, flashing the central pattern – a calligraphic motif of the name of God – like the wings of a butterfly basking in sunshine.

Then, after commenting on how the usual Western reflection on Islamic art 'fails to answer the question of meaning behind it', he comes to the heart of the experience:

Something was getting under my skin as my eyes roamed the walls. I had a feeling that this was different from any art I had ever seen. And in that cold, lowering dusk, in that shabby courtyard, where the tilework is a third destroyed, a ray of meaning seemed to leap from the walls. It was as if they had suddenly become articulate and, shedding for a moment their almost formal precision, began to dance and weave with meaning. It was the mathematics of it, just like the geometrical precision of atoms in a crystal, that lent them such force. From every panel, every frieze, burst an expression of the same creative breath, each an encrypted fragment of the Divine. This was not the art of decoration but of sacred ciphers, in which the onlooker is invited to participate, not merely stand in awe, and decode the patterns according to his means.

Here the description catches the appear*ance* of meaning. This is active, working meaning, coming to life in understanding, not finished meaning which is present as an object to a subject. Here we have the unitary event of {meaning/understanding} in which there is no separation between meaning and understanding. Far from being just a subjective experience, this is a 'non-Cartesian' event which happens upstream *before* the separation into subject and object. Consequently it is not surprising that in such an event there is no longer any separation between inside and outside, so that what is usually experienced as being 'inside' may appear to come from 'outside'. In this hermeneutic reversal, understanding is in-formed by the meaning – so that the 'subject' really does become a *subject* for the meaning. This is the reverse of the usual notion that, in understanding, the subject 'grasps' the meaning.

Much has been said about what Gadamer meant by his famous statement that 'Being that can be understood is language', but near the end of his life he said that what he meant is 'Being that can be understood begins to speak to us'.[19] This seems to be just what is happening in the hermeneutic reversal described by Jason Elliot. In such an encounter we are not subjects actively projecting meaning into things, but instead we become receptive subjects for the meaning which appears. This reversal of the subject – which returns us to the original meaning of 'subject' before Kant inverted it – is the condition for the possibility of understanding (adopting Kant's language). The subject becomes a subject for the meaning, instead of the source of meaning as it is usually portrayed as being in modern philosophy. Goethe expresses the reversal beautifully:

> I do not rest until I have found a pregnant point, from which much can be deduced, or rather, that freely brings forth much out of itself and bears it towards me, since in working and perceiving I proceed carefully and faithfully.[20]

There is a reversal of intentionality here – what we could call a 'counter-intentionality'. If, as Heidegger says, 'intentionality' means 'directing-itself-toward', then in the hermeneutic reversal there is clearly this sense of 'directing-itself toward', but in this case it is coming-toward and constituting the subject, instead of going-toward and constituting the object.[21] In Aristotle's language, a text has the potential to mean and a reader has the potential to understand. There is a *single* actualisation of both, which can be described either as the actualising of meaning or the actualising of understanding, because the actualising of meaning *is* the actualising of understanding. As in the case of the teacher and the student discussed previously, we can see that here the actualising of meaning (i.e. mean*ing*) occurs in the reader. But this certainly does not make it *just* subjective, i.e. something which belongs only to the subject, since clearly it is the meaning itself which actualises in the reader. We could say that the meaning participates the reader. We would usually put this the other way round and say that the reader participates the meaning. But the difficulty with doing so is that, almost without our noticing, this makes us into an active subject whereas we are in fact receptive. If we participate the meaning, it is because primarily the meaning participates us – and *this* is understanding. In the hermeneutic reversal the meaning *becomes* us – which is not at all the same as saying that the meaning becomes *us*. We could put it the other way round, and say that we are *becomed* by the meaning, and in this sense we become a subject for the meaning.

This reversal is very much in tune with phenomenology, as Richard Palmer makes clear:

> The combination of *phainesthai* and *logos*, then, as phenomenology means letting things become manifest as what they are, without forcing our own categories on them. It means a reversal of direction from that one is accustomed to: it is not we who point to things; rather, things show themselves to us. This is not to suggest some primitive animism but the recognition that the very essence of true understanding is that of being led by the power of the thing to manifest itself ... Phenomenology is a means of being led by the phenomenon through a way of access genuinely belonging to it.

He goes on to emphasise the significance which this has for hermeneutics:

> Such a method would be of highest significance to hermeneutical theory, since it implies that interpretation [understanding] is not grounded in human consciousness and human categories but in the manifestness of the thing encountered, the reality that comes to meet us.[22]

It is evident that such understanding will require a change in attitude on our part. When we describe the event of understanding we usually do so from a subject-centred perspective – as when we say that we have 'grasped the meaning'. This perspective emphasises the role of the subject as being active rather than receptive in understanding. Of course, in order to understand the subject has to engage herself actively with the work which is to be understood. But this does not mean that the event of understanding is itself primarily the act of a self-centred subject. When understanding happens, the subject becomes receptive rather than active. When we fail to notice this reversal, we tend to think of understanding in terms of either assimilation or appropriation. When we eat, the food we ingest must be made similar to our own body – this is the process of assimilation. As applied to understanding the meaning of a work, this is the process of taking something new, and therefore different, and coming to see it in terms of what is already familiar to us, i.e. making it similar to what we know already. Appropriation, on the other hand, goes beyond this basic nutritional approach to understanding in that it makes use *in its own way* of what it finds in the text. In appropriation, the text is not reduced to what is known already, but is used by the reader in the service of her own interests. So instead of the text being used to consolidate what we understand already, it is used to expand our understanding further by adding to it whilst remaining within the same overall horizon. In other words, we can make use of it in our own way by incorporating it into what concerns us. We make it our own, so that it is no longer just something left over from the past which has to be reconstructed in the present, but which is used by being accommodated to the present in order to enlarge our understanding of our own interests. Thus appropriation is a way of including something new in understanding, but in such a way that once again a subject-centred approach is emphasised. In appropriation the subject makes the meaning her own, without reducing it to what she already understands (which would be assimilation), but she does so only in a way that expands rather than transforms her understanding. In other words, in appropriation the self-centred subject controls the use to which the meaning is put, and hence understanding is under the control of the subject.

Understanding which participates meaning clearly goes beyond both assimilation and appropriation. In this case we find ourselves being addressed by the text and experience a reversal in the direction of meaning over which we have no control. This is no longer a subject-centred experience, but one in which the subject is transformed by the encounter with meaning instead of using it for her own purposes. This usually begins with a failure to understand. We are 'pulled up short by the text', as Gadamer puts it, when we feel that we cannot understand it, or that it seems to be saying something unexpected.[23] We do not understand in a vacuum. We always already understand, and it is this already-understanding that is 'pulled up short' by the text and found to be inadequate. The text calls our already-understanding into question, with the effect that, when the meaning of the work participates us, our understanding is transformed – not consolidated or expanded – so that we understand differently. For this transformation to take place we have to become open to the text, and this openness becomes possible as a consequence of the experience of failing to understand. This is the condition for the hermeneutic reversal in which the meaning participates us to transform our understanding, so that in the hermeneutic encounter 'the reader is not so much the interpreter as the interpreted'.[24] In the event of understanding, therefore, it is not so much we who appropriate the meaning, but we ourselves who are appropriated by the meaning of the work. So we are participated by the meaning that we participate in – this is the hermeneutic reversal.

The Problem of Multiple Meaning

So far we have proceeded as if the meaning of a work which comes into being in the event of understanding is simply univocal, i.e. that it is the same on all occasions of its appearance. However, this fails to take into account the fact that different people, at different times, in different situations, see differences in the meaning of a work and therefore understand it differently. How are we to understand this? If we think in the framework of the subject-object dichotomy, then we can only conclude that these differences in understanding must be subjective. If we think that the meaning of a text is what was in the author's mind, then we must think of the meaning as a finished object which is self-contained and unchanging. In this case, understanding must happen when the meaning that the author had in mind is reproduced in the mind of the reader, and hence any differences in meaning must be entirely subjective and not part of the objective meaning of the text as originally intended by the author. Understanding is therefore entirely reproductive, and it must exclude what now seem to be misunderstandings, so that we can come closer to the supposed, single true meaning of the text. But if these differences in meaning between different interpreters are conceived as being misunderstandings, then the question arises of how can we know which meaning corresponds to what the author had in mind? One way to try to salvage something here is to introduce a distinction between the meaning of a text and its significance. We have looked at this already, and we saw that this distinction itself rests squarely on the subject-object separation, and consequently only leads us into the dichotomy of objectivism and relativism.

There is another possibility if we think in the framework of the subject-object separation, which is to say that, no matter what the author may have had in mind, the meaning of the text is now simply whatever subjective response arises in the mind of the reader. In which case understanding is certainly not reproductive. The reader's understanding no longer supposedly reproduces what the author had in mind, but is simply whatever subjective experience is occasioned in the reader's mind by his or her experience of reading the text. So instead of one meaning there will be many. It no longer even makes sense to refer to the 'meaning of the text', since there are now many meanings. We have swung from the pole of objectivism to the opposite pole of relativism. But if the meaning of a work were to be determined exclusively by its reception by the reader, this would have the consequence that each occasion of understanding would result in the production of an entirely new work. One way of understanding a work would be no more, or no less, legitimate than any other – resulting in what Gadamer calls 'an untenable hermeneutic nihilism'.[25] The work would in effect fragment into many different works.

The problem of the dichotomy of objectivism and relativism has used up a river of ink in philosophy – especially in the last decades of the twentieth century. But all to no avail, because no resolution of this dichotomy is possible within the framework of the subject-object separation, since this is the source of the dichotomy in the first place. So the question is how to understand the fact that there are differences in the meaning of a work when we begin, not with meaning as a finished object, but with the coming-into-being of meaning in the unitary event of {meaning/understanding}? We have seen that, when we do this, we find that 'meaning is always *coming into being* through the "happening" of understanding', so that "understanding must be conceived as a part of the process of the coming into being of meaning'.[26] Understanding is no longer conceived as the *duplication* of the meaning that the author originally intended. But if the meaning we understand is not objective in this sense, then neither is it only subjective – in which case the work would fragment into many meanings and we would collapse into relativism. Instead of being either objective or relative in this sense, understanding is 'the concretion of meaning itself', so that meaning *comes into being* in understanding. Thus the meaning of the text is actualised and comes-to-presence in understanding *now,* instead of being a replication of what had once been present 'in the author's mind'. So the differences in the meaning of a work which appear in different situations, on the different occasions of its actualisation, must therefore belong to the work itself as actualisations of the *work's* own possibilities of meaning. Instead of this being the fragmentation of the work into a sheer multiplicity of meanings, these differences are 'the work's own possibilities of being that emerge as the work explicates itself, as it were, in the variety of its aspects'.[27] So instead of there being only a single meaning, or alternatively no more that what Gadamer calls 'a mere subjective variety of conceptions', there is the meaning of the work be-ing itself multiply as differences of itself. This clearly requires a way of thinking that:

> ... will explain, first, the fact of multiple interpretation; second, that multiple interpretations can all be true to the work; and third, that the work can be multiply interpreted, multiply true, without disintegrating into fragments or degenerating into an empty form. It is a formidable task.[28]

It is indeed! What is needed here is a new idea of 'the one and the many' in which, instead of either/or, it is possible for the very same thing to be both one and many at the same time.

The One and the Many

It is when we shift our attention into the lived experience of meaning – the happening of meaning which is the event of understanding – that we find ourselves encountering the need for a new idea of 'the one and the many' which is at first unfamiliar. We find that, instead of the dichotomy of 'one' and 'many', the meaning of a work can be one and yet multiple – multiple without being many – so that there can be multiple meaning without this becoming many meanings. Now this is just what we have seen with Goethe's dynamical thinking of 'the one and the many' in the life of the plant. So here we find an *organic* idea of the one and the many which will help us to see how there can indeed be multiple interpretations that are all true to the work, so that 'the work can be multiply interpreted, multiply true, without disintegrating into fragments or degenerating into an empty form'. In terms of the notion of the intensive dimension of One, which we introduced in the previous chapter, we can say there is One meaning, without this being one meaning (the error of objectivism), or the multiple modes in which this meaning manifests becoming many different meanings (the error of relativism). The meaning of a work can be intensively multiple without becoming extensively many meanings.

We can begin to see the difference between the intensive and extensive forms of 'the one and the many' quite easily in the case of the performing arts, such as drama and music. If we go to see *Hamlet,* for example, it is Hamlet that we see. This seems so obvious that we wonder why anyone would even mention it. But consider the following 'advertisements' for the performance (see Figure 12).

The Bard's Players
Present
HAMLET
By
William Shakespeare
Tonight at 7.30

The Bard's Players
Present
A Reproduction of HAMLET
By
William Shakespeare
Tonight at 7.30

Figure 12. Advertising a performance of Hamlet.

The first would give us no concern, but the second would seem decidedly odd. In fact we would feel that it was a mistake, although we may not at first realise that becoming clear about the difference between the two would take us on an adventure into the one and the many.

Where does *Hamlet* exist? We might say that once it existed in the mind of Shakespeare and now it exists in the script. But this is no more than ink on the page until it is brought to life. We might read it to ourselves, or aloud with others, but we would soon realise that this is not enough because *Hamlet* only exists fully when it is performed. As Gadamer says, 'we must recognise that "presentation" is the mode of being of the work of art'.[29] So if the work of art has its being in presentation, the performance is not an optional extra but the event in which *Hamlet* comes into being and exists fully – we might say that the performance is the doing which be's *Hamlet*. The words 'presentation' and 'present' are especially appropriate here. As well as the idea of 'putting on a performance', they also connote the idea of 'making present' in the sense of 'bringing into appearance' and thereby 'bringing into the present'. These all give us the dynamic sense of an event. So, when we go to see *Hamlet* it is *Hamlet* that we see because *Hamlet* comes-to-presence and is present in the present-ation.

Evidently, if the work lives in its presentation, then it cannot be separated from its presentation. So we cannot have the work-in-itself *and* its presentation. There is no 'pure' *Hamlet* apart from its presentation, and yet each presentation is an interpretation. But because the work lives in its presentation, it follows that each presentation is an interpretation of the work itself and not merely a subjective interpretation imposed upon it. In other words, the interpretation *belongs to the work* even though it can come-to-presence only through the actors, the audience, and the director. Weinsheimer expresses this very clearly (with a change of play):

> When I go to the theatre, what I see there is an interpretation of, say, *Macbeth;* but I also see *Macbeth* itself. There is nowhere I could go to find the uninterpreted *Macbeth* because interpretation brings it into existence. But I do not also see *Macbeth*, as if seeing the play itself were distinct from seeing the interpretation. The interpretation is *Macbeth* itself and vice versa.[30]

The work and its interpretation are not two different things that can be separated. If we become aware that what we are seeing is an interpretation, this in itself is an indication that the work and the so-called interpretation have become differentiated, and hence that in this case the interpretation does not belong to the work, and is therefore not in fact an authentic interpretation. In such a case it is no more than the director's subjective viewpoint which has been superimposed upon the work, interfering with its integrity, and resulting in something which is not a genuine interpretation of the work but only masquerades as one, and should therefore be more properly called a pseudo-interpretation. It is the experience of such counterfeit interpretations, which *are* differentiated from the work, that helps us to understand that there are interpretations which are authentic in that they are interpretations of the work itself – the work's own interpretations which belong to the possibility of the work:

> There are in principle, then, true and adequate
> interpretations, whose truth consists in the fact that they are
> not distinguishable from the work itself. True interpretations
> are interpretations *of the work itself.* They are to be explained
> therefore by reference not to the meaning-conferring acts of
> the interpreter but to the work itself.[31]

In a good performance, all that goes into it – the acting, staging, lighting, direction, etc. – becomes transparent, so that it is the play itself which is present. There will be many such present-ations of *Hamlet* , but this would not lead us to say that there are many '*Hamlets*':

> On the contrary, one and the same *Hamlet* exists in the
> many performances. *Hamlet* itself is contemporaneous with
> them all, despite the fact that each performance may be
> significantly different as well. In fact, Gadamer wants to say
> that it is precisely the type of being which a play like *Hamlet*
> has that enables it to be what it is as a temporal reality whose
> being is always in some sense different in each of its discrete
> presentations, while at the same time maintaining its identity
> throughout them all.[32]

We can begin to see here the intensive form of 'the one and the many' which we have encountered previously. If for the moment we ignore the significant differences between productions, we can recognise that, whereas there are extensively many performances, there is intensively One *Hamlet*. Because it is *Hamlet* itself which comes-to-presence on each occasion when it is presented, and not a reproduction of *Hamlet,* instead of many '*Hamlets*' there is One *Hamlet* in the 'multiplicity in unity' of its presentations. Because the coming-into-being of *Hamlet* is therefore literally occasion-al, or episodic, it would be easy to miss the intensive form of 'the one and the many' here and, instead of the 'multiplicity in unity' of the intensive dimension of One, just see many separate performances extensively. This would be equivalent to missing the organic unity of the King Edward potato plant as it grows through time (Chapter 3), and instead thinking of this intensive 'multiplicity in unity' as just many separate potatoes.

Now we need to take into account the fact that there will always be significant differences between presentations of a play, i.e. there will be real differences in interpretation and hence in understanding. Where such differences are involved – differences which belong to the work itself and are not just subjective 'additions' to it – we have to go beyond just multiplicity to genuine diversity. We explored this in the previous chapter in connection with Goethe's dynamic idea of 'the one and the many' in the metamorphosis of the plant. There we discovered the key idea of *self-difference,* which is clearly intensive, so that 'multiplicity in unity' now becomes the dynamic unity of self-difference. As we will see, this dynamic mode of unity is found in the life of meaning as well as in organic life. However, at this point we are going to return to the question of the multiple meaning of written texts, instead of focusing on the performing arts (we focused on this first because in this case it is quite easy to see the intensive/extensive distinction in the idea of 'the one and the many'). But in order to do this, we must first say something about language – which will be a preview of the next chapter.

The Act of Saying

Contrary to the widespread view, language is not primarily an instrument by means of which our thoughts are expressed in words, as if these were no more than clothes in which thoughts are dressed for public presentation. It may be that language is almost (but never quite) reduced to an instrumental function in the millions of documents that are routinely produced every day in business and government – just think of the output from the European Commission or the United Nations. But what we are concerned with here is at the opposite pole to this frenzy of linguistic activity. We are concerned with texts (which are far fewer in number) that are expressions of original thinking. Such texts might be creative works of literature, or any of the original works of philosophy from Plato onwards, where it is well-known that differences in meaning appear each time the work is read. In such original works language is nothing like an instrument just 'putting into words' what has already been thought 'in the author's mind'. On the contrary, here language is the medium in which thinking comes into expression – in which thinking forms itself into expressed thought. When we approach it dynamically, we see that the language in which thinking comes into expression gives form to *what* is thought, and does not take what is already thought and simply dress it in words. Yet at the same time this does not mean that what is thought is only the product of the words which express it. Gadamer says that 'a thought first attains determinate existence in being formulated in words', commenting on which Wachterhausser says:

> This means that the thought is neither found pre-existing the words, which would make the words a mere 'casing' for thought, nor is it the case that the words themselves produce the thought such that the thought is 'mere words'.[33]

Thus saying is not a case of just embodying meaning which is already determinate, but of bringing it into determinate form through its expression. If what is expressed is not fully determinate beforehand, but only becomes so in being expressed, there is no *separation* between what is expressed and its expression – and yet this does not mean that what is expressed is *reduced* to its expression. Here we need to think in a way that does not separate into two, but at the same time doesn't collapse into one.

So when we try to understand the meaning that comes-to-expression in the text, we have to enter into the dynamic condition of saying. We have to go 'upstream' from *what* is said into the *saying* of what is said. This is the *event* of saying which is the coming-into-word. Now we begin to become aware that what is said in the text is expressed within a context of which we are not immediately aware because our attention is focused on *what* is said. This con-text is 'everything meant "with" or by a text, but which remains unspoken'.[34] What is said does not encapsulate its own meaning, as if it stood on its own and could be understood entirely independently of what is not said but is also meant along with it (the context). On the contrary, what is not said forms the background within which what is said emerges and upon which it depends for its meaning. What is said always carries with it a 'circle of the unexpressed', which means that it always 'carries within it unspoken meanings and possibilities of understanding'.[35] Consequently, when we try to understand a work we can do so only by entering into the *language* of the text and trying to 'read between the lines' –which does not mean interpolating our own thoughts into the text. This means that to understand the work we not only have to attend to what is said in the text, but also become aware of what is not said but which is nevertheless the context within which what is said comes into expression, and which is therefore the unexpressed dimension of meaning which comes through what is expressed. What is said always means more than is expressed, so 'there exists no statement which one can grasp only from the content which it presents'.[36] Every word that is said 'carries with it the unsaid', which Gadamer emphasises is not an imperfection of its expressive power, but an expression of 'the living virtuality of speech, that brings a totality of meaning into play, without being able to express it totally'.[37]

We can understand this in terms of the act of distinction, which was discussed in the first chapter. There we saw that a distinction is a unitary act of {differencing/relating}, which becomes evident when we try to catch distinction 'in the act' instead of beginning with the outcome. If we begin with *what* is distinguished, we lose sight of the intrinsic relation in the distinction, so that it seems as if a distinction only separates. In the present case, the act of saying brings meaning into expression, and in doing so creates a distinction between expressed and unexpressed meaning. The key difference here is that meaning which is expressed *appears* – it appears *as appearing* – whereas meaning which is not expressed doesn't appear. In terms of the act of distinction, the meaning which is expressed does not stand on its own, self-contained, but is related internally within the act of distinction to what is not expressed. Thus the expressed meaning is always within the context of what is *thereby* unexpressed. Consequently, instead of being self-contained, the expressed meaning must always be open to being understood differently – where the difference belongs to the meaning and is not simply a subjective projection. The key to understanding this is to go 'upstream' into the act of saying. This is a unitary act of distinction, which in this case takes the form {expressed meaning/ unexpressed meaning}. In this way we can see clearly that there *must* always be more in the act of say*ing* than can be expressed.

Self-Differences of Meaning

Now we can see the origin of the differences in meaning that different readers see in one and the same work. It is clear from what has just been said about language, that the meaning of an original text cannot be something that is finished. Such texts are charged with a virtuality of meaning which belongs to the work, but which can be actualised only in the event of understanding. As we have seen, if meaning comes into being through the happening of understanding, then understanding must be conceived as the coming into being of meaning. So if different readers understand one and the same work differently, then the differences must be differences in the coming into being of the meaning of the work. They are therefore self-differences of the work itself, and not different subjective interpretations which readers impose on it. David Linge says of these differences in the meaning of a work:

> They are not alien or secondary to it but are its very being, as possibilities that flow from it and are included in it as facets of its own disclosure. The variety of performances or interpretations are not simply subjective variations of meaning locked in subjectivity, but belong instead to the ontological possibility of the work.[38]

Gadamer says of this variety in interpretation:

> Thus it is not at all a question of a mere subjective variety of conceptions, but the work's own possibilities of being that emerge as the work explicates itself, as it were, in the variety of its aspects.[39]

Which Weinsheimer expresses concisely: 'The work is the multiple possibilities of its interpretation.'[40]

Gadamer insists that the work itself has meaning that can be understood. But this meaning is not simply 'there' in the text because it comes to realisation only in the event of understanding (we must remember that this is prior to the subject-object separation). Consequently, as conditions and circumstances change – personal, social, cultural-historical, and so on – the work will be understood differently. But these differences belong to the *possibility* of the meaning of the work itself:

... it is not the case that the work exists *an sich* and only the effect varies: it is the work itself that displays itself under various conditions. The viewer [reader] of today not only sees things in a different way, he sees different things.[41]

We can see that, in the case of written texts, this is a consequence of the inevitable fact that what is said always 'carries with it the unsaid', which is the background context within which what is expressed *means*. Since what is said, therefore, does not (cannot) have a self-contained meaning, the meaning of a work can come into being differently because different conditions and circumstances will elicit different possibilities of *its* meaning. Thus 'it remains the same work whose fullness of meaning is realised in the changing process of understanding', and 'even if it must be understood in different ways, it is still the same text presenting itself to us in these different ways'.[42]

These variations which belong to the work itself are clearly *self-differences*, so that the unity here, which is the unity of coming-into-being, is the dynamical unity of self-differencing. Whereas extensively these differences can only appear as a symptom of the work falling apart into many meanings, intensively we can see that they are in fact the dynamic unity of the work itself. This is clearly indicated in the first quotation above, where these differences are said to 'flow from it [the work] and are included in it as facets of its own disclosure'. The sense of the language here is clearly that the differences belong to the work itself and are not 'a mere subjective variety' – cf. also 'the work explicates itself ... in the variety of its aspects' and 'it is the work itself that displays itself under various conditions'. However, we must be very careful to shift our focus 'upstream'. Otherwise we can all too easily fall into the trap of reading what is said here in a semi-static way – 'facets' and 'aspects' can (although they need not) encourage this – and as a consequence we begin to think of 'possibilities' as if they were already included in the work in a 'downstream' manner. Thus we think of possibilities as if they were actualities-in-waiting which have not yet emerged, but which are ready to do so when the conditions are right – which is a case of the kind of back to front thinking that 'tries to reach the milk by way of the cheese'. Instead we need to think dynamically here, and it will help us to do so if we now return to the dynamic idea of 'the one and the many' that we saw in the living plant.

117

Organic Hermeneutics

In the previous chapter we saw how Goethe came to the intrinsically dynamic notion of the self-differencing plant organ, ever changing into other modes of itself without becoming other than itself. What looks at first like a different organ is in fact the same organ differently and not another organ – it becomes 'other' without becoming 'another'. By becoming the other of itself (an intensive distinction), the metamorphosing organ can become multiple without becoming many (which would be an extensive distinction). So here 'the one and the many' takes the intensive form of 'multiplicity in unity' – a form in which there can be multiplicity within unity without the unity being fragmented thereby into a mere multiplicity. In the previous chapter, we saw how the philosopher of Goethean science, Ron Brady, characterised the intrinsically dynamic form of life as 'becoming other in order to remain itself', and the quotation which we gave there will bear repeating in the present context:

> The forms of life are not 'finished work' but always forms *becoming*, and their 'potency to be otherwise' is an immediate aspect of their internal constitution... The *becoming* that belongs to this constitution is not a process that finishes when it reaches a certain goal but a condition of existence – a necessity to change in order to remain the same.[43]

We can recognise what is said here about organic life reflected in what has been said about understanding the meaning of a work. The meaning of a text is akin to a form of life in that it is not 'finished work' but always becoming. Also, a 'potency to be otherwise' is evidently an immediate aspect of the internal constitution of a text, as we see from the fact that differences in meaning are described as 'possibilities that flow from it and are included in it as facets of its own disclosure'. We have seen that this is a direct consequence of the way that the meaning of a text is both actual and virtual at the same time on account of the fact that what is said can never include its meaning totally within itself. So the differences in meaning are self-differences of the meaning of the work itself, which emerge when it is understood in a variety of contexts as the conditions and circumstances change. In other words, these self-differences constitute the dynamic unity of the work itself, and not the disintegration of the work into many different meanings. The meaning comes-into-being differently in different situations, but it *is* the meaning of the work itself and 'not simply subjective variations of a meaning locked in subjectivity'. Thus the 'organism' of the work is an inexhaustible 'multiplicity in unity' of self-differences, which are the works own possibility of meaning manifesting in a variety of contexts and situations.

In the organic case, we saw how different environmental circumstances *evoke* the potential of the plant to express itself in a form which is appropriate to the specific conditions. It was emphasised that this does not mean that the environment *determines* the specific form which the plant takes. The plant is not passive but active in responding to the challenge of the environment, because it is a *living* organism and not an inert body. The conditions influence the specific form which the plant manifests, but they do not cause it externally in what would be a mechanical way. The living organism produces itself actively, instead of being conditioned passively, in response to the environment. Thus the plant responds actively out of its own 'potency to be otherwise' – 'becoming other in order to remain itself' – to express the form of itself which the environment elicits. We can see something equivalent to this in the presentation of a work – whether it be the reading of a written text, the presentation of a play, or the performance of a piece of music. Thus, with a text, the differences in meaning which manifest in different contexts and situations are not merely subjective variations imposed on the text, but expressions of 'the possibility of meaning' of the work itself. But we also saw that we must guard against falling into the opposite trap by thinking of the specific form which a plant takes as if it were predetermined by the organism itself. On this view, the different forms which are observed are already there, as if stored in the plant waiting to emerge, and when the external conditions are right the corresponding form will emerge. But this is not at all the case. On the contrary, the form which the plant takes in given circumstances is a concrete expression of the dynamic possibility of the plant, not a pre-formed possibility which merely comes out when circumstances permit. In the case of a text, the equivalent of this error would be to think of the self-differences in the meaning of a work as if they were already predetermined in the text itself. But this is 'finished product' thinking. In any particular situation, the dynamical 'possibility of meaning' of the work is evoked in accordance with the conditions of that situation, but the meaning which thus comes into expression is a manifestation of the possibility of the work which is in accordance with that specific situation, and is not a preformed possibility which is already there in the text. The difference here is between an 'upstream' dynamic approach, and one which is 'downstream' and thinks in terms of preformed possibilities which effectively 'puts the cheese back into the milk'.

So when Weinsheimer says that 'the work is the multiple possibilities of its interpretation', this is intended to be understood *dynamically* and not as referring to predetermined possibilities. Shelley gives a graphic image of this in his *Defense of Poetry*:

> A great poem is a fountain forever overflowing with the
> waters of wisdom and delight; and after one person and one
> age has exhausted all its divine effluence which their peculiar
> relations enable them to share, another and yet another
> succeeds, and new relations are ever developed, the source of
> unforeseen and unconceived delight.[44]

This gives us a very different picture from the view that a poem has a single meaning which it is the task of interpretation to 'unveil' – or indeed from the alternative view which would reduce the meaning of a poem to whatever the reader finds in it, as if the meaning belongs to the reader more than it does to the poem. This is the familiar opposition between objectivism and relativism, where in the case of the latter the poem effectively splinters into as many subjectively different 'poems' as there are readers. But far from being, either a finished meaning to be 'unveiled', or 'a mere subjective variety of conceptions', the image of meaning which Shelley gives is one which is totally dynamic. The gushing up of the fountain, forever overflowing, gives us an image of meaning coming-into-being inexhaustibly. The fact that there is no single correct interpretation to be unveiled does not require us to go to the other extreme of supposing that the meaning is somehow undecidable. The dynamical approach to hermeneutics frees us from the constraint of believing that the meaning is *either* determined *or* undecidable, by showing us that it is in fact inexhaustible. The poem – or indeed any literary or philosophical text – is a cornucopia of meaning ever coming-into-being as different modes of itself. The renewal of meaning in each new situation is not another, different meaning, but the self-differencing of the meaning of the work itself.

If we think of the Peony, with its thousand different varieties, we have a striking image of this cornucopia of meaning. The Peony gives us an astonishing picture of the plant's 'potency to be otherwise' which it expresses in 'becoming other in order to remain itself'. So the variety, which we see extensively as many different plants, is organically One plant coming-into-being intensively as different modes of itself. This means that the diversity we see *is* the dynamic unity of the plant. So if we were able to go to the Chelsea Flower Show on the day when the varieties of Peony are exhibited, what we would see in front of us would be unity in the 'disguise' of diversity. Since we do not usually recognise this unity which is 'hidden' as diversity, we go looking for it instead in a way that seeks to reduce the diversity to what is common, and thus end up with uniformity. If we now look at the differences in the meaning of a work in this organic way, we can recognise that, what at first seems to be the work fragmenting into many different meanings, is in fact the dynamic unity of the work itself. Since the different interpretations are self-differences of the work itself, the diversity of interpretations *is* the dynamic unity of the work and not its fragmentation. Once again, the unity is in front of us, where we don't expect to find it, in the 'disguise' of diversity. How different this living understanding is from any misguided attempt to find unity of meaning in the diversity of interpretations by looking for what they have in common. All the arguments and hand wringing about relativism (which should not be confused with relativity) disappear when we consider meaning in the manner of the dynamic form of life. Does the horticulturalist throw up her hands in horror because there are a thousand varieties of Peony? No, she rejoices in it. Nobody complains about 'relativism' among the peonies. Nobody asks how we could find out which one is 'the true Peony', because they are all the true Peony. We would not think of calling the Peony 'postmodern', or declaring that there is nihilism at the heart of the Peony. The case of the organic shows us how we can understand the diversity of interpretations without raising the 'spectre of relativism' – of which Gadamer has so often been wrongly accused.[45]

The Enhancement of Being

The differences in the meaning of a work which appear in different situations are now clearly seen as belonging to the work itself. Furthermore, when we see this in the light of the dynamic idea of 'the one and the many', which we discovered in Goethe's organic thinking,

121

then we can also see that this does *not* imply relativism because these differences are self-differencings of the meaning of the work itself coming into being. Since this variety of interpretations – which is in fact the unity of the work – can happen only in the context of different situations, then evidently the work needs these differences in conditions and circumstances to enable its potential for meaning to be increasingly realised. If we imagine the artificial case of a text which was always understood in the context of the very same circumstances and conditions, we can see that this would really be an impoverishment of the work, even though at first we might have the comforting illusion that this restriction of the work's 'possibility of meaning' would represent the definitive understanding. In order to become itself more fully, the work needs the variety of interpretations which different situations make possible. So instead of this leading to an intractable relativism – as we would expect it to do if we were 'downstream' in the subject-object framework – we can now see that this variety is necessary for the coming-into-being of the meaning of the work itself.

Gadamer refers to this as an 'increase in being' of the work:

> Far from signalling a depletion of an original meaning or the destruction of its identity, Gadamer says that the process of one meaning unfolding its many historical possibilities actually signals an 'increase in being' (*Seinszuwachs*).[46]

We might think at first that these differences in meaning which emerge with time would have the effect of undermining the work, and hence of diminishing rather than enhancing it. But the very opposite is the case. Far from multiple interpretation depleting the meaning of the work – as if there were some original, self-contained 'pure' meaning – it enhances the meaning, because with time different possibilities of the work's meaning can come into manifestation. Thus the work becomes itself more fully with each manifestation – we could say that the meaning of the work 'grows' with interpretation in different contexts – so that the work's reality is increased with each event of understanding. The meaning of the work can only be realised over time (though never finished), not because 'temporal distance' is needed for us to get the meaning of the work 'into perspective', but because it is time which brings the new situations and contexts within which the possibility of the work's meaning can come into expression more fully so that the work 'increases in being'.[47]

We can readily recognise the phenomenon of enhancement or 'increase in being' in living nature. If we take the Rose, for example, and consider the difference between the natural and cultivated forms in which it occurs, we can see quite clearly that the Rose is 'increased in being' by its cultivation. It is important to remember here that any cultivated form is not something which is imposed on the Rose by the breeder, but is always a form which it is possible for the Rose to take – even though it may not necessarily do so of its own accord without the breeder's artificial selection. The thousand varieties of Peony, to which we have referred, is clearly another example of 'increase in being' in the organic world. Yet another is provided by the variety of forms which the pigeon can take as a result of the breeder's art – and which fascinated Charles Darwin so much. The diversity here can readily be seen as the enhancement of the pigeon when we compare it with the basic form of the naturally occurring rock pigeon, from which all the 'fancy' forms ultimately originate.[48] In none of these examples would we look upon the rich diversity of forms as indicating a depletion or dilution of an original 'pure' organism which gradually loses its identity. On the contrary, we see this diversity as the organism's way of becoming itself more fully. But in the case of meaning, instead of seeing that it is through its multiple interpretations that a work 'comes into its own', we emit cries of despair that we will ever be able to understand the meaning of the work, and consequently face the horror of sliding into relativism. However, we are not dealing with the 'finished meaning' of the work, but with the unfinished meaning which is always coming-into-being in understanding, so that, far from being diminished by multiple interpretations, the meaning of the work comes more fully into its own.

Gadamer points to other instances where there is a greater degree of realisation as a consequence of interpretation in different conditions and circumstances. In the law, for example, we might think at first that the law is a universal under which all individual cases are subsumed, so that it is just a matter of applying the law directly to each individual case, i.e. that it will be applied in the same way to all cases. But matters do not turn out this way in practice. When it is applied to an individual case, the law itself has to be interpreted in the light of that case. This does not mean modifying the law, and it certainly does not mean changing it into another one. It means that the law is not simply a general principle under which all individual cases are subsumed. Rather, the way in which

123

the law manifests in a particular case makes the law itself manifest, which means that, not only is the individual case clarified by the law, but the law itself is clarified in turn by the individual case to which it is applied. The *possibility* of the law in question is therefore brought out by the specific circumstances. But of course we must not think of this possibility as if it were already there, determined beforehand – which would be 'finished product' thinking. We must think *dynamically* of the law coming-into-being differently in its application to different individual cases. What this means is that the law 'comes into its own' more fully through its application:

> Talk of the law 'in itself' apart from this historical process
> of concrete interpretations would be nonsense. But as a
> temporal reality the law evolves, which does not mean that
> every new application gives us a new law but one and the
> same law is 'increased' over time.[49]

Gadamer sees this legal case as itself a specific instance – an individual case – of the hermeneutic problem, which he relates to the wider philosophical problem of 'the universal and the particular' that is familiar from medieval philosophy:

> If the heart of the hermeneutic problem is that one and
> the same tradition must time and again be understood in
> a different way, the problem, logically speaking, concerns
> the relationship between the universal and the particular.
> Understanding, then, is a special case of applying something
> universal to a particular situation.[50]

We are familiar with the idea of 'the universal and the particular' in mathematical thinking, and also with the idea of universal laws of nature, and as a consequence we tend to think of the relationship between universal and particular in a unilateral way. The effect of this is that we can all too easily *separate* the universal from the particular, thereby introducing a false dualism. Thus, in this style of thinking, the universal determines the particular, which is therefore subsumed under the universal. For example, every possible triangle is subsumed in advance under the universal concept 'triangle', of which any triangle is therefore a particular instance. Everything is included in the universal concept, so it is unthinkable that the universal itself could be enhanced by any *particular* triangle. The movement is only from the universal to the particular and never the other way round, so there simply cannot be any enhancement of the universal by the individual case. Thus when Gadamer says understanding is 'a special case of applying something universal to a particular situation', he clearly does not have in mind the unilateral case with which we are so familiar from the mathematical style of thinking – and which has entered into our customary way of thinking more than we may recognise. In contrast to this, the relationship between the universal and the particular in hermeneutic thinking is not unilateral, because in this case the universal itself is reciprocally determined by the individual case to which it is applied. So in this case the particular contributes to the universal, which therefore cannot be understood in advance of its application to individual cases. This does not mean that something extra is just added on to the universal, or that it becomes a new and different universal, but that each individual case to which the universal is applied thereby contributes reciprocally to the enhancement of the universal so that it comes more 'into its own'– which is what is meant by 'increase in being'. We are now so accustomed to the universal in the mathematical style of thinking, that at first it seems as if the universal in hermeneutics isn't really a universal at all. But it should be the other way round. We should look upon the unilateral relationship between the universal and the particular in mathematics (as well as in the kind of philosophy that looks to mathematics for its model) as being only a special, restricted instance. We should certainly not look upon this as the ideal against which all other instances should be judged, in which case they can only appear to represent a weakening of the universal and not its enhancement.

We can recognise the reciprocal movement between universal and particular in the case of understanding texts, as well as in the application of the law.[51] Gadamer sees 'application' as an integral feature of understanding, not something that takes place *after* understanding has happened. He says 'understanding is always already application'.[52] At first we may find this quite difficult to grasp because of our ingrained habit of thinking that application is subsequent to understanding, not an integral part of it, and hence that we can first understand a text *per se* and then afterwards use it for particular applications. But Gadamer makes it clear: 'Application does not mean first understanding a given universal in itself and then afterwards applying it to a concrete case', because 'application is neither a subsequent nor merely an occasional part of the phenomenon of understanding, but codetermines it as a whole from the beginning'.[53] In traditional hermeneutics it was thought that understanding is immediate, what just happened automatically, and that interpretation is needed only on the occasions when there is some obstruction to understanding so that it is no longer immediate. This separation of understanding from interpretation has been superseded – especially since Heidegger – and it is now recognised that 'Understanding is never immediate but always mediated by interpretation; and since this is always the case, understanding is indivisible from interpretation'.[54] But application was still seen as being separate from and subsequent to understanding. Now Gadamer has also removed this separation 'by regarding not only understanding and interpretation, but also application as comprising one unified process', so that in the hermeneutics of unfinished meaning, 'application is integral with, and indivisible from, interpretive understanding'.[55] This is the holistic form of the dynamics of the coming-into-being of meaning as understanding. We can now see why the meaning of a work only begins with the author, and must always be ever unfinished but never incomplete. Understanding is always an adventure in meaning. It is not about trying to find our way back into the past, as if the text were no more than a memorial. We become involved with what the text says, 'sharing in what the text shares with us,' so that understanding is 'not a repetition of something past but the sharing of a present meaning'.[56] In this way the life of meaning continues, so that the work is one with the history of its understanding, to which we now each make our own contribution as we also try to understand – provided we always remember that 'the interpretive activity considers itself wholly bound by the meaning of the text'.[57]

The dynamic unity which manifests as the diversity of interpretations of the work is what Gadamer calls the work's effective history – we remember here that the work cannot be *separated* from its interpretations. What this means is that the work becomes its own tradition. The idea of tradition is often misunderstood, usually in a way that gives it a conservative function which is quite contrary to the *openness* of hermeneutics – which is concerned with *unfinished* meaning. It does not just mean that the work has a place in a tradition along with other works, but that the work itself *becomes its own tradition*. So when a work is referred to as a 'traditionary text', what this means primarily is that the work cannot be considered apart from the history of its interpretation. The work as we encounter it now carries its effective history with it. Each interpretation is an expression of the meaning of the work and the 'multiplicity in unity' which the work thus becomes is the dynamic unity of the self-differencing of the work. The differences are the work's differences; they belong to the possibility of the work and are not just imposed on it externally. The 'traditionary text' is therefore the unfinished meaning of the work ever coming-into-being as a dynamic whole.[58]

5. Catching Saying in the Act

Everyone is familiar with the experience of sitting down to write something, and finding that as pen is put to paper (or fingers to keyboard) the words just seem to escape us. It is very frustrating. We feel that we know what we want to say, until we try to say it, and then to our dismay we find, not only that we cannot say it, but that *what* we want to say no longer seems as clear to us as we thought it was. It seems to have withdrawn, as if we can't quite see it, even though at first we felt that we knew what it was we wanted to say. This experience is as disconcerting as it is familiar. But, sooner or later the words come, and as they do we see clearly *what* it is we want to say. What had tantalisingly eluded us, as if somehow just out of reach, now comes into expression as we recognise 'that says it'. We must attend to this very carefully if we are to see what is really happening here. At this point, for example, it is only too easy to fall into the trap of subjectivism, and think that 'expression' here means that the subject is expressing what he or she has 'in mind'. This is a fallacy. When the words come it is what is meant that comes into expression and thus appears – when we can say it we see it. It is not primarily the writer expressing herself, but the thing meant. What is said, the content, is more than the words in which it is said, and yet without the words what is meant would not appear. The words do not produce what we say, as if what is meant is a product of the words, but they do bring it into appearance. This is why it is so important to find the right words, i.e. the words which do express what is meant so that it appears.

The important point here is that it is not just a matter of the words 'merely making plain what is being thought of beforehand', but that 'a thought first attains determinate existence in being formulated in words'.[1] But we often get this backwards, and so miss the point:

Though we are sometimes inclined to say that we cannot find words adequate to express our thoughts, reflection on such situations often seems to point to the conclusion that, until we find words to express a particular thought, the thought itself is vague and indeterminate.[2]

Coming-into-language is the fulfilment of thought. As Merleau-Ponty puts it: 'thought tends towards expression as towards its completion', so that expression 'does not translate ready-made thought, but accomplishes it'.[3]

But it is what is meant that is seen through the language which expresses it, and not the language itself, which seems to become transparent as we see the meaning through it which otherwise would not be seen. This is the extraordinary thing about language. It is the medium in which what is meant appears – and without which it would not appear – and consequently like any medium, such as the air in which we live, it is in a sense 'transparent' because we 'see through it' to what is meant:

Even when we are talking about language, our own words recede from view and hide themselves precisely in expressing the thing meant. Language is most itself when it is transparent and self-concealing, when it reveals not itself but its object.[4]

Because language disappears in its function and is thus self-concealing, we only become aware of language as such when there is a 'breakdown' – as when we can't find the words to say what we mean and consequently feel that we just can't see it. The word that says it and what is meant belong together in this unitary language event. The words are not added to the meaning, but nor do they produce the meaning – as if this were no more than a product of the words. On the contrary, the words and the meaning belong together in such an intimate way that it is through the words that the meaning appears. But when it does, it is the meaning itself that comes into expression, not just the words.

Understanding language is like walking along a tightrope. We can so easily lose our balance and fall off on one side or the other. On one side we fall into the fallacy of believing that words simply express thoughts which are already formed – in which case language would be no more than the clothing of thought, merely its 'casing' or the 'wrappings' in which it is packed.[5] On the other side we fall into the fallacy of believing that the words produce the thought, as if it were merely the words. One way we underestimate language; the other way we overestimate it. Either way, we miss language.

It is not difficult to see how each of these misunderstandings arises. Because we see the meaning manifest in language, and it is the meaning that we see and not the language itself, we can quite easily wrongly conclude that we can have unmediated access to 'pure' meaning which we first know non-linguistically before words are subsequently superadded for the purpose of communication. In this case the function of language seems to be entirely secondary. The 'pure' meaning is *embodied* in language, a process which does not influence the meaning itself, which therefore can subsequently be released from its linguistic embodiment into the understanding of the reader or listener, where it will again become a 'pure' meaning. We can see that language is 'too late' here. It implies that the meaning is already formed beforehand, and consequently it overlooks the formative role which language has in making the meaning manifest – 'But to talk of "making manifest" doesn't imply that what is so revealed was already fully formulated beforehand'.[6] We can only avoid being 'too late' by trying to catch language 'in the act' of saying. This fallacy of the preformation of the meaning underestimates the role of language because it begins 'downstream' with what has already been said, instead of going 'upstream' into the *saying* of what is said. This is why it is just a bit 'too late', and consequently back-projects the end result – i.e. the expressed meaning – prior to the event of coming-into-language. We can now also see how the opposite misunderstanding arises when, noticing that we only come to the meaning through language, we wrongly conclude from this that the meaning itself is entirely a *product* of language, i.e. that 'the words themselves produce the thought such that the thought is "mere words"'.[7] In this case language is 'too soon', introduced prior to the meaning in such a way that it can only seem as if the meaning itself is no more than the words which express it. When we are 'too soon', it seems that language actually creates the meaning in the first place instead of bringing it into expression. This linguistic reductionism assumes that thought is *only* words. But it is no more 'only words' than it is 'pure meaning'.

We need to think in a way that does not separate into two but at the same time doesn't collapse into one. If we can walk along this tightrope without losing our balance, we discover how words and meaning are intimately linked in the act of saying in such a way that words neither create meaning, nor merely reflect meaning that is already formed. We see the meaning through the words that say it: 'Language has the capacity to point to something that is not a product of language, but which is nevertheless always grasped by linguistic means'.[8] This is remarkable, because it flatly contradicts our common sense assumption that words are just tools, merely conventional signs, for representing what we have already understood. The reason why this instrumental view of language isn't true is expressed very clearly by David Mitchell:

> It is certainly misleading to think of language in general as a tool or instrument, at any rate when the purpose of our speech is to express that which we claim to be true. I choose one tool rather than another to achieve a particular result; for example, I use a fork, it may be, rather than a spade to dig my garden. It would be misleading to describe as a tool anything the operation of which is not a matter of choice or selection on my part. Thus, although I may select one medium of communication rather than another, in that I may decide to express myself in writing rather than orally, and although I may choose one word in preference to another, it is not the case that I decide to use 'language-in-general' to express my thoughts. Thought realises itself, is actualised, only in language.[9]

This why Gadamer – as well as Heidegger and Wittgenstein – understands language as the medium in which meaning becomes manifest and thus appears, instead of considering language as being primarily a tool or instrument of subjectivity. Language lets things come into meaning so that they can be understood. We can understand this by shifting the position of attention in experience 'upstream', from the word that is already said into the dynamic word coming into being in the event of saying.

Disclosure and Representation

If we begin at the end, with what is said, it seems as if the words just *represent* what is there already. Our everyday experience confirms this. We say 'there is a glass of water on the table', and this draws attention to the situation that is there. The sentence does this by representing what is there, in the double sense of functioning as a representative for the state of affairs, and also of re-presenting it. This is the familiar function of language. But it is not the only function, as we have seen above in the case where what we mean first appears in saying it. In this case it would not be correct to say that it is 'there' already, and saying it just represents it. In such a case 'expression is no longer simply inert', as it would be if 'the content precedes its external means of expression'.[10] When we shift the position of attention within experience 'upstream' from what is said into the saying of what is said, we discover that language does not simply represent what is already there, but on the contrary it brings it into expression so that it becomes 'there' in the first place. Only when something is *already* present can we talk about language representing it. What easily gets overlooked is the way that it comes to be present in the first place, so that it can then be represented. When we say that something – an entity, a state of affairs – is re-presented in language, this presupposes that what is re-presented is already present – i.e. that it has already appeared and is manifest. What this overlooks is the way that is through coming into expression in language that it *appears* and is 'there' in the first place. What is represented is *first* disclosed through language which *then* represents it. We have to think dynamically here, otherwise we begin at the end with *what* is said – the outcome – and thereby miss the primary disclosive function of language on which the secondary representational function depends. When we go 'downstream' from the appear*ance* to what appears – from the *saying* of what is said to *what* is said – then automatically it seems that language just represents what is there, whereas it is through language that it is 'there' in the first place. This shift is inevitable, but it can be reversed.

Two things need to be noticed here. Firstly, disclosure is obscured by virtue of its own nature. Attention is focused on *what* is disclosed – on *what* appears – and hence the event of disclosure which is the appear*ance* is overlooked as a consequence of the event itself. We could say that the event of disclosure is self-concealing. So, paradoxically, as a result of its disclosive function, it seems obvious to us that language is representational. Because we focus 'downstream' on *what* is disclosed, and hence miss the disclosive event itself, it seems obvious to us that words represent what seems to be just 'there' and ready for words to be superadded. The second thing which needs to be noticed is that, whereas representation clearly fits the subject-object separation, disclosure does not. In the representational mode of language, what is designated by the words is conceived as being present already as an object, which is then represented in words by a subject. But it is quite different with the disclosive mode of language. In this case the object is not there in advance of saying it, as if it were already present as an object. What is said appears in the act of saying it – it is constituted (but not produced) in the act that says it – and is not present already as an object that can be re-presented in words by the subject. In its secondary representational mode language fits the subject-object model of experience, but what we can easily overlook is the way that in its primary disclosive mode language is prior to the subject-object separation. This dichotomy does not occur 'in the act' but only subsequently with the outcome. In fact this is always true for experience *as it is lived*. The subject-object separation comes in only with the reflection of experience after it has been lived – although at that stage we mistake the reflection for lived experience itself. Thinking dynamically here enables us to see that these are two different phases of experience. In the previous chapter we have seen the way that understanding the meaning of a text also doesn't fit the subject-object model of experience. The unitary event of {meaning/understanding} is prior to this dichotomy, because the happening of understanding *is* the coming into being of meaning. We saw that this is a specific instance of the unitary event of {appearing/seeing}, and we can recognise now that the event of disclosure has the same structure.

The disclosive function of language is sometimes revealed very clearly in cases where there is a 'breakdown' in the normal development of language in a person. An exceptional example of this is given by the remarkable story of Helen Keller.

As a very young girl, Helen Keller had a severe attack of measles, which left her deaf and blind. This happened to her before the dawning of language, and it was only due to the extraordinary work of her dedicated governess that these extreme difficulties were eventually overcome. The moment when this finally happened is described in her own words:

> We walked down to the well-house, attracted by the fragrance of the honeysuckle with which it was covered. Someone was drawing water and my teacher placed my hand under the spout. As the cool stream gushed over one hand she spelled into the other the word 'water', first slowly then rapidly. I stood still, my whole attention fixed upon the motion of her fingers. Suddenly I felt a misty consciousness as of something forgotten – a thrill of returning thought; and somehow the mystery of language was revealed to me. I knew then that 'w-a-t-e-r' meant the wonderful something that flowed over my hand. That living word awakened my soul, gave it light, joy, set it free! ... I left the well-house eager to learn. Everything had a name, and each name gave birth to a new thought. As we returned to the house each object that I touched seemed to quiver with life. That was because I saw everything with the strange new light that had come to me.[11]

She is blind but describes herself as *seeing* with a new *light*. The word 'water' does not represent or stand for water here; it is not a label to be attached to water for the purpose of communicating information. Helen Keller does not already know water, to which she then adds the word. No, in this case everything is reversed. The word 'water' *shows* her water; it brings it to light so that she *sees* it. Here the name calls water into appearance; it calls water into being as *water*, instead of the indistinct sense awareness which there had been before. (We should note that the first few sentences in the quotation describe the situation before the dawn of language as it could only appear to her after language had dawned in her. This is inevitable, because she is giving an account, but we must consciously allow for it.) Thus the word here is not a sign in the sense that it designates something already known, because the thing designated by it would first have had to be seen independently of language – and evidently it had not been. The word 'water' allows water to manifest, and in this sense the word can be said to indicate 'water' – i.e. it is the word itself that indicates 'water', not Helen Keller who indicates 'water' by means of the word. One of the major obstacles to catching language in the act is what can be called the 'myth of subjectivity', which emphasises self-consciousness and consequently conceives the individual self as being the centre and origin to which everything must be referred. This leads us to believe, quite wrongly as it happens, that language is primarily an instrument of human subjectivity – which would mean in this case that it was Helen Keller who indicated by means of the word 'water' that it was water, and clearly this gets it back to front.

To catch language 'in the act' we have to go back 'upstream' from the said into the say*ing*, so that we enter into the primary event of language as disclosure instead of remaining with the secondary function of language as representation. This is what Heidegger is referring to when he says that *'the essential being of language is Saying as Showing'*, and that 'saying is in no way the linguistic expression added to the phenomena after they have appeared'.[12] A sign is to be understood fundamentally as 'showing in the sense of bringing something to light'. Heidegger emphasises the transformation which takes place when we do not understand the sign in this way, but think of it instead as something that designates. When this happens, 'the kinship of Showing with what it shows' is lost and becomes 'transformed into a conventional relation between a sign and its signification'. When Heidegger says that the essence of language is 'Saying as Showing', he does not mean Saying to be taken in the sense of a being that says, but more in the sense of a Saying which 'be's'. Similarly with Showing: this is not showing a being – like shining a light on an object in the dark – but the Showing which is its appear*ance* wherein it 'be's'. In Helen Keller's experience, the word 'water' *says* water in the sense that it *shows* water (*not* points to it), whereby *water* appears. The word does not designate water *after* it has first appeared. But after water has appeared we take it that this is what the word does, because we are now 'downstream' at the stage where it seems that the word is separate from the thing, and consequently that the word is no more than a label we attach to the thing – so that language becomes merely representational. But language is primarily the disclosive event of *saying-showing-seeing*. These are not component parts but holistic aspects: saying *is* showing, and showing *is* seeing because there could not be *showing* without seeing – cf. the discussion of appearing and seeing, meaning and understanding, in the previous chapter. This is the 'upstream' disclosive event of language.

Primarily the word indicates in the sense of 'shows'; it does not designate in the sense of 'points to'. We must remember here that it is the word that indicates, and not the individual subject that uses the word to indicate – in which case it would be the subject that indicated, and not the word. But the word can be used secondarily to designate (point to) something which it has primarily disclosed. For example, if the word 'water' has shown water, so that water manifests, then the word can be used to designate water in the sense of drawing attention to water – as in 'look at the water over there', or 'mind the water on the floor'. What usually happens is that we only recognise the secondary representational word, which designates, and not the primary disclosive word which is 'the condition for the possibility' of the word being used secondarily to designate. The disclosive word which shows makes it possible for the word to designate – we could say that the disclosive word is the background within which it is possible for the word to be used to designate. We don't notice this disclosive background, without which it would be impossible to use language to designate, because it is upstream from the secondary function of language upon which our attention is usually focused.

We can see from this how easily we can underestimate the role of language by mistaking what is only secondary for what is primary. On the other hand, it is also possible to overestimate the role of language. Although this distortion is less common than the former, nevertheless we do sometimes find this kind of exaggeration among philosophers. For example, in an interview on television in 1978 with Bryan Magee, the Oxford philosopher, A.J. Ayer, said that: 'the world is the world as we describe it'. In his later philosophical autobiography, *Confessions of a Philosopher*, Magee says why he disagrees with this:

> If I look up from the writing of this sentence, my view immediately takes in half a room containing scores if not hundreds of multicoloured items and shapes in higgledy-piggledy relationships with one another. I see it all clearly and distinctly, instantly and effortlessly. There is no conceivable form of words into which this simple, unitary act of vision can be put ... Even something as simple and everyday as the sight of a towel dropped on to a bathroom floor is inaccessible to language – and inaccessible to it from many points of view at the same time: no words to describe the shape it has fallen into, no words to describe the degrees of shading in its colours, no words to describe the differentials of shadow in its folds, no words to describe its spatial relationships to all the other objects in the bathroom. I see all these things at once with great precision and definiteness, with clarity and certainty, and in all their complexity, and yet I would be totally unable, as would anyone else, to put that experience into words.[13]

Magee concludes that it is emphatically not the case that, as Ayer said, 'the world is the world as we describe it'. If it were so, we could describe the taste of boiled potato in such a way that anyone who had not tasted it would know from the description what it tasted like! On the contrary, 'all that language can do is to indicate with the utmost generality and in the broadest and crudest terms what it is that I see'.[14] This is certainly correct, and yet the way it is expressed here does not reflect the sheer miraculousness of seeing 'what it is' that Helen Keller experienced, which is clearly anything but a case of 'all that language can do'.

The word 'water' did far more for Helen Keller than 'indicate with the utmost generality and in the broadest and crudest of terms'. On the contrary, she tells us, 'That living word awakened my soul, gave it light, joy, set it free!' so that everything she touched 'seemed to quiver with life' because she 'saw everything in the strange new light' that had dawned in her. This ontological event is the light of language which brings into appearance what things are. Of course, once we become familiar with it, we no longer experience language in this way. We simply get on with living in the world in which language makes it possible for us to live – the everyday world of human life. We cannot help but 'forget' the ontological event of language, and begin to think of language in a more mundane way, and even try to give a naturalistic explanation of language in terms of what is not language. All of this, inevitable as it is, takes us away from language *as* language, so that now, in order to understand language, we need to be able to recover in some way the *experience* of language itself – which is what Heidegger means by 'the way to language':

> Instead of explaining language in terms of one thing or
> another, and thus running away from it, the way to language
> intends to let language be experienced as language.[15]

The word 'water' does not describe water – that is impossible. The word *says* water, which means it *shows* water so that water appears as such in Helen Keller's experience of *seeing* that it *is* water. We could simply say that water manifests – but we have to be careful to catch this 'in the act' of manifesting, and avoid the pitfall of nominalism which thinks of water as manifest already and then named 'water'. Manifesting means 'the coming into appearance of the intelligibility (*logos*) of what is' – which means that 'the intelligible is not simply copied in language', but that 'in language the intelligible forms itself'.[16] So 'what comes into language is not simply an object of state or affairs in the "external" world, but something in its understandableness, its meaning'.[17] This is just what is expressed so succinctly in Gadamer's much quoted remark that 'being that can be understood is language'.[18]

The event of manifestation in which things appear as what they are – that water *is* water – is a unitary event of {appearing/seeing}. In Heidegger's sense this is the 'to be' of things which otherwise would just exist:

> When the carrot shoots emerge from the earth, they display themselves for the rabbit as well as for the gardener. The difference is that the gardener *understands* that the carrots *are* carrots; he can be aware of their manifestness as such in a way the rabbit cannot.[19]

It is this be-ing that we miss when we focus attention on the appearance of what has already appeared instead of its appear*ance*. The appear*ance* of things is concealed by the things that appear – it is as though the event of manifestation withdraws in the 'draw of things' which are thereby manifest and thus occupy our attention.[20] What Helen Keller experienced was the be-ing (the 'to be') of water, which we overlook when we fail to distinguish be-ing from being, in which case we consider water (or a carrot) only as a substantial thing – as a being but not be-ing. When we focus on being, language can only seem to be representational in function; but when we shift from being to be-ing, we discover the primary mode of language as disclosure.

The distinction between disclosure and representation enables us to understand what poetry is more clearly. We often just take it for granted that poetry is different from any other way of using language. We feel that there is something special about poetry. Well it is special, but it is not unique:

> It is as though poetry above all discloses the secret truth of all literary writing: that form is constitutive of content and not just a reflection of it.[21]

Poetry is language which is purely disclosive – a poem cannot represent anything. If we haven't recognised the dynamic difference between disclosure and representation, and think that language is primarily representational, then we cannot help but conclude that poetry is in some way different from any other use of language. But in fact the disclosive power of poetry is really a *heightened* manifestation of what is primary to *all* language. Far from showing us something different about language, what it actually does is to make visible what is fundamental to language itself – and which the secondary mode of language as representation is dependent upon:

> So it is as though poetry grants us the actual experience
> of seeing the meaning take shape as a practice, rather than
> handling it simply as a finished object.[22]

It is when we see the articulated meaning as a finished object that the word disappears in the meaning – because we see *through* language to the meaning – and language as the medium in which we experience the meaning becomes invisible to us. Since we now believe that we have the meaning without language (because it is transparent to us), it must seem that the role of language is to represent (not disclose) the meaning. A *separation* between word and meaning is consequently introduced where there is none in the event of disclosure. What poetry does is not only to give us an exemplary case of language as disclosure, but it does so in a way that negates the transparency of language by drawing our attention to the words themselves. We are invited to become aware of and enjoy the experience of the words in the act of disclosing the meaning they constitute.

Introduction into Language

We think that we give the meaning to words, whereas in the first place it is words that give meaning to us. The confusion here arises because we don't distinguish clearly enough between learning to speak language in the first place, and further languages which subsequently we may also learn to speak. We do not give sufficient attention to the significance of the fact that, when we learn a second language, we can do so only because we have already learned to hear and speak language. The child learns whichever language she hears being spoken in her own surroundings. If it is English, she learns to understand and speak English, whereas if it is Chinese she will begin to speak and understand Chinese. But we must be careful how we describe this, otherwise something significant here will easily be overlooked. We might just say that the child is learning to speak English, or learning to speak Chinese. But if we do, we are describing it in the same way that we would if the child were learning English, or Chinese, as a second language. But in this case the child has *already* acquired language, i.e. the possibility of learning a second language presupposes that first the child acquires *language* as such. They are not the same thing. In the primary case the child doesn't learn to speak Chinese, as we would do secondarily, but rather it learns to *speak* and it does so 'chinesely'. The child first learns to speak 'chinesely', or 'englishly', or in whatever mode, and only after that would it be appropriate to say that the child is learning to speak Chinese, or English, if it is doing so as a second language.

The original dawning of language is a unique event in our lives. For each of us it only happens once in this way, and any further language we may then learn is done so already within the horizon of language, and hence cannot be a guide to understanding language itself. Thus we can learn that 'acqua' means the same as 'water' – or that 'water' means the same as 'acqua' – because we already have the idea from language in the first place. So now we can give the meaning to the word – i.e. we give the meaning 'water', which was given to us in the first place by the word 'water', to the word 'acqua', which therefore can now mean 'water' for us. For an Italian speaker learning English it would simply be the other way round. This procedure for assigning meanings, when learning a second language, gives us the sense of the word as being a sign, i.e. that basically it 'stands for' or 'represents' something because this is the meaning we have given to it. But when we think that a word is basically a sign, what we have done is to mistakenly transpose what is secondary into what is primary by imagining that this is what the word 'water', for example, does for us *in the first place.* What this overlooks is that, when we say 'acqua' is the word for water, we are only able to do so because we already have the concept 'water' which was given to us by language in the first place. Overlooking this, we forget that we cannot recognise something independently of its concept, and so we easily imagine that when we see water we are doing so 'directly', i.e. without any concept and therefore independently of language – which therefore, it seems, can be added on later. Georg Kühlewind describes this very clearly:

The awakening of our first language occurs very differently from the learning of a second language. Learning a second language is a dualistic process because we have already been given the meaning in our (first) mother tongue, and we then merely learn the more or less corresponding expression in the second language. The first language *creates* the meanings that are then 'named' in the second language. In fact, this process reinforces the impression that the world is built up nominalistically because we easily forget that we can perceive a thing only if it already has a meaning, only if it is already defined by a concept. *Before* the first language or mother tongue, there is nothing that could be named.[23]

The word is not primarily a sign because 'it is not an existent thing that one picks up and gives an ideality of meaning in order to make another being visible through it'.[24] If it were like this, then it would be a tool which we made, and thus language would be an instrument used by a subject to organise and thereby exercise control over an object. We have already seen that language is not primarily an instrument of human subjectivity – the case of Helen Keller shows this very clearly – although it can be used secondarily in this way. Language is more like a medium within which we exist than an instrument which we use. We do not give 'an ideality of meaning' to the word, because 'the ideality of the meaning lies in the word itself. It is meaningful already'.[25] This is the extraordinary thing about language, which we so easily overlook because we are always trying to explain language in terms of something other than language. If the word is already meaningful, then it means itself and we cannot refer it to something other than itself that bestows meaning on it. But if the word *means*, then concomitantly it must be something that can be understood, because 'meaning' *necessarily* entails 'understanding'. So, for example, the meaning of 'water' *is* understanding 'water'. Now we have seen something like this already in the previous chapter. There we found that, when the position of attention is shifted 'upstream' into the *event* of understanding, instead of being focused 'downstream' on *what* is understood, no separation is possible between meaning and understanding because the event of understanding *is* the appear*ance* of meaning. There is a single event, the unitary event of {meaning/understanding}, which is a special case of the unitary event of {appearing/seeing}. Now we can recognise the same event

141

with language itself, and so discover the fundamental reason why we find it in the case of understanding the meaning of a text.

Language has its own mode of being – although the way we usually think of language always attributes another, non-language kind of being to it. We have seen the difficulty that arises as a result of 'limiting the possibilities for understanding language to the sort of being that belongs to a sign', instead of 'considering that language could have its own mode of being' which is different from that of a sign.[26] What is this unique mode of being? From all that has been indicated above we can answer that the being of language is saying – i.e. that language *is* saying; that language *says* – which brings us back to Heidegger. We don't experience the say*ing* of language when we have become adults, but we did when we were young children. When language first dawns in us we are upstream in the coming-into-being of language. There, like Helen Keller, we experience language 'in the act' instead of as something which is already a finished product. What is encountered in this 'upstream' experience is the unique mode of being which is the *saying* of language: 'Language is unique in that it is not just perception but *meaningful* perception. Children must grasp both perception and meaning *at the same time*'.[27]

To paraphrase Schelling, we need to shift the focus of attention from language as fact to language as 'the action itself in its acting'. This is the difference between language as product and language as productivity – what we could call 'language languaged' and 'language language-ing'.[28] When we go upstream into the coming-into-being of language in this way, what we encounter is language be-ing itself differently. So whereas downstream we find many different languages, what we encounter upstream is better described dynamically as the self-differencing of language. We recognise here the dynamic idea of 'the one and the many' which we first explored in detail in connection with Goethe's way of seeing the living plant – learning 'to think like a plant lives' – and which we then went on to see exemplified in the dynamics of understanding the meaning of a text in Gadamer's hermeneutics. All that we have found previously in this regard also turns out to be the case for language itself, provided that we enter into the dynamic phase of language, instead of focusing downstream on language at its finished stage. If we don't distinguish carefully between first learning language, and subsequently learning a second language, then we will think of language more as a fact, as something which is already a finished product, instead of catching language coming-into-being. It is at the stage of the finished product that we find many languages, whereas in the dynamic phase, instead of *separate* languages, we find language be-ing itself differently. So we find self-different modes of language (intensively) instead of many different languages (extensively) – in Deleuze's expression: 'there is *other* without there being *several*' (see Figure 13).[29]

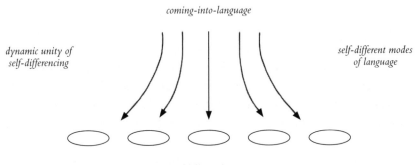

Figure 13. Coming-into language.

It is well known that young children who are beginning to acquire language can, if the environmental circumstances are right, acquire more than one language at the same time. The facility with which they do this is quite remarkable. They can switch from one to the other effortlessly, but without confusing the two – by switching in mid-sentence, for example. It is clear that in this case the child is *not* learning a second language, i.e. a *separate* language, as we would do as adults. The child is in the 'upstream' dynamic phase of learning language, *before* the state of separation, so that here she can come-into-language in different modes of itself. Although it may not seem this way to adults, for the child here there are self-different modes of language instead of separate languages. Consequently it is possible for the child to come into language 'chinesely' as well as 'englishly', say, and hence she can quite easily switch from one mode of saying into the other. It is later, when the dynamic phase of learning language has passed, that the now older child will learn a second language as *another* language. Then she will have to learn a second language as a finished product, whereas in the dynamic phase she can enter into the self-differencing of language.

The diversity which we see as many different languages is really the dynamic unity of language itself. It is language be-ing intensively multiple instead of extensively many – it is 'multiplicity in unity' instead of 'many ones'. So the diversity of language, far from being merely the fragmentation of language into a plurality of many languages, is in fact the dynamic unity of the coming-into-being of language itself. By learning to think dynamically we can come to recognise diversity as unity, and thereby reciprocally to recognise unity in the form of diversity. We have seen this intensive form of 'the one and the many' in the life of the plant as this is described by Goethe. Thus we saw how, in the case of the Peony, each of the many different varieties is the self-differencing of the plant, so that the diversity we see *is* the unity of the Peony. In the same way, in Gadamer's hermeneutics, we came to recognise the diversity in understanding as self-differences in the meaning of the work itself, so that the diversity of interpretations *is* the dynamic unity of the work. So now we find this intensive form of 'the one and the many' in the self- differences of language – which are more usually thought of as just different languages. This diversity which is the unity of language could be called the efflorescence of language. This may be more apt than first appears, as Mark Abley shows:

> The greatest concentrations of linguistic diversity are found in the tropics. Languages thrive where the web of biological diversity is also at its most intricate: in tropical rain forests, above all ... The wet, hilly, verdant island of New Guinea ... has given birth to about eleven hundred living languages. One in every six languages spoken on the planet comes from this island. (These are languages, remember, not dialects.) In New Guinea each valley, each mountain, each tributary, each bay seems to have a language of its own. Five times more languages are native to the island than to the entire continent of Europe. New Guinea is also the hotbed of biological diversity on a scale almost inconceivable in colder realms.[30]

It follows from the dynamic approach that the diversity of the modes of language is as much to be expected as the diversity of life itself. This diversity is therefore 'natural' to language, and not a deficiency which needs to be overcome. Merleau-Ponty says 'there is only one language in a state of becoming', and that if we 'renounce the abstract universality of a rational grammar which would give us the common essence of all languages', we will discover the concrete universality of language 'which is becoming different from itself while remaining the same'.[31]

Language and World

The word 'language' is a noun, and its use can easily seduce us into thinking that there is a separate entity called language, which exists as such apart from and independently of whatever is said in it. Although we can and do consider 'language' in abstraction from everything that can be said in it – as if it were a pure form, i.e. purified of content – we must recognise that this *is* an abstraction, and that, although it may be a useful fiction for some purposes, it is nevertheless artificial and can only lead us astray if we take it to be real. What is real is *living* language, and this cannot be separated from whatever is said in it, so there can be no *separation* of language from what is said which considers language 'as an object in itself and apart from what it means' – which is the artificial separation that is made in linguistic science.[32] We can see why it is unrealistic to make this separation by considering what has been said above about the experience of language as *saying*. If language *is* saying (the be-ing of language is saying), then clearly language cannot be considered apart from *saying*. But, since there cannot be saying without something being said (saying must say something in order to be 'saying'), then language cannot be separated from whatever is said in it – in whatever mode this may be, i.e. 'englishly', 'chinesely', and so on. 'Language can be considered as an object... only when it is idling and abstracted from its use.... in use, language is always saying something'.[33] This usually does not occur to us, and we think of language as if it existed in itself entirely independently of what is said in it. This is the form-content dualism which, like all dualisms, arises because our thinking begins 'too late' – in this case, too far 'downstream' to catch language in the act of saying. It is this dichotomy which 'is entailed by the initial move that institutes linguistic science: the creation of language as an object in itself and apart from what it means'.[34] Heidegger contrasts this technical-scientific comprehension of language, which attempts to grasp it as a formal system, with what he calls the *hermeneutic* experience of language, which takes us 'upstream' into the saying of language.[35]

Language is the medium in which things can *appear* as such, i.e. as *what* they are. Thus water appears as 'water' – but it is water which appears, not the word 'water'. The word calls water into appearance as 'water', and in doing so brings it into the world. When things enter into language they enter the world. What appears in saying are things

145

themselves – language is the medium, not the message. This also applies to language itself. Language appears as such, i.e. as 'language', only on account of language, so that language is always more than can be encompassed by 'language'. Since it is language which gives us 'language', it is more like the horizon which is the condition for there to be a space in which things can be, but which cannot itself ever be within that space. When we treat language as a formal system we are considering it as something we discover in the world, without noticing that it is language which gives us the world in the first place – i.e. that language is the condition for the possibility of there being 'world'. The world 'lights up' in the dawning of language. When Helen Keller says 'I saw everything with the strange new light that had come to me', this is the light of intelligibility in which what things *are* appears. Language brings us into the world in which *we* live – which means the world in which 'we' lives. Language articulates 'our world'. This does not mean that the world is produced by language, but that, as Gadamer puts it, 'in language the world itself presents itself'.[36]

We usually think of the world as the totality of objects – the sum total of all the entities there are – and we think of language as a cloak of representation thrown over the world. This is how it seems in the framework of the subject-object separation. But this is not how it seems when we approach it phenomenologically. In this case the world is not the totality of entities but the context within which entities can appear. It is one of Heidegger's remarkable achievements to have rescued the world from the epistemological approach of the Cartesian tradition and bring it back into lived experience.[37] In fact nothing could be more familiar to us – or more overlooked – than 'world', because it means the world we are always already 'in'. This primary existential meaning of 'world' is expressed very clearly by Timothy Clark:

> We bring with us, even in the simplest kind of task or
> statement, a sense of 'world'. 'World' is one of the major
> terms in Heidegger's thinking, in the early work often close
> in meaning to 'being'. It means no particular entity (it is
> not the planet or the globe itself) but is that presupposed
> and disregarded space of familiarity and recognition within
> which all the beings around us show themselves, *are* for us.
> That is to say, Heidegger's concept of 'world' is close to the
> common meaning of the term when we talk about 'the world'
> of the Bible, or the 'world' of the modern Chinese or modern

English – i.e. the fundamental understanding within which individual things, people, history, texts, buildings, projects cohere together within a shared horizon of significance, purposes and connotations. One might use the term 'world-view', but this falsely suggests that a 'world' is a particular stance that people or individuals hold inside their heads, as representations, rather than the more fundamental shared disclosure of things within which they find themselves in all their thoughts, practices and beliefs, providing the basis even of their self-conceptions and suppositions.[38]

The world in which we live is not in the first place a world without language, to which language is subsequently added, as if 'words are subsequently joined to something already known, prior to its verbalization'.[39] But then equally, we do not first have language and then try to discover what it corresponds to in the world. In the first case we would have to associate words with things, whereas in the second we would have to find things to fit the words. These are equally unsatisfactory, and in fact they really both come down to the same thing: the *separation* of language and the world. Although we may talk about 'language and the world', the 'and' is fictitious because it implies that we could have 'language' and 'world' separately. But in fact we cannot, even though we are accustomed to thinking as if we could. Gadamer, following Heidegger's lead, emphasises that 'language is not just one of man's possessions in the world; rather, on it depends the fact that man has a *world* at all'.[40] It may not seem so at first. For example, at this moment I can see a glass of water on the table in front of me, and it seems clear to me that I don't need language to see this. I don't have to pronounce inwardly to myself the sentence 'there is a glass of water on the table in front of me' in order to see that there is a glass of water on the table in front of me. However, whereas I certainly don't need to do this, I do need the concepts which enable me to see the glass of water on the table, and these concepts are given to me through language in the first place. If the gift of language had been withheld from me I simply would not be able to see this – no matter how well-developed my sensory-motor capacity had become through the experience of manipulating material bodies as a child (which is an indispensable basis for language). But it goes both ways: if we cannot have the world without language, then reciprocally we cannot have language without the world. This may seem surprising at first, but it is the other side of the 'and': if we cannot have the world *and* language, as if they were two, concomitantly we cannot have language *and* the world. In the first case the world would come first, independently of language; in the second case language would come first, independently of the world. But what would language be like if it were independent of the world? What would it *say*? As Gadamer puts it:

> Language has no independent life apart from the world that comes to language within it. Not only is the world world only insofar as it comes into language, but language, too, has its real being only in the fact that the world is presented in it.[41]

If language and the world belong together, we should try to avoid speaking about 'language and the world', or even about 'the world', and instead refer to 'the language-world'. This does not mean that the world is only language (which it isn't), nor does it mean the world is described by language (we have seen that it can't be). It is meant to indicate that the world comes into being within the medium of language – which is *not* the same as saying that the world is language. But equally we must avoid thinking of the world as if it were already there beforehand, and then comes into language, i.e. as if language is only a dressing in which the world is presented *for us*. There is no such pure, undressed world-as-it-is-in-itself. We often do think in this way, but it is 'downstream' after-the-fact thinking. It introduces a fictitious dualism whereby we believe that we can conceive of a 'world in itself' which is beyond language, and from which language therefore effectively cuts us off. So now there is a supposed 'world in itself' which is 'behind' language, and which can only appear to us through the 'veil' of language. Hence the world-as-it-is-in-itself is separate from its appearance-for-us, and so we can only have a 'worldview' which is subjective. We are now back in the subject-object dichotomy. But this forgets what the world is, and replaces it with a supposed world-object which we can never know as it is in itself but only in our representation of it. However, the world cannot be seen as an object because it is prior to the separation into subject and object. On the contrary, as we have seen, instead of language coming between us and the world, we find that 'in language the world itself presents itself'.[42] Language is not a filter through which we can only ever have a view of the world as an appearance-for-us which is different from the world-in-itself: 'Rather, what the world is is not different from the view(s) in which it presents itself'.[43]

It only looks like language *and* the world – as if they exist independently and are brought together extensively – when we begin downstream with the world already languaged. But if we shift upstream to try to catch language in the act, then we find, not just that language discloses world, but that language and world are disclosed *together*. The 'language-world' is really the concrete phenomenon, from which 'language on its own without the world', and 'world on its own without language' are abstractions. When the child first learns to speak, it is not language on its own that she acquires, as if she first acquired language and then applied it to the world. If this were the case, she would need to have an already intelligible world prior to language, and language itself would be an empty form – it would be world-emptied:

> We don't first disclose a totality-of-significations in the world and then go on to map language onto it. But neither is it the case that we first discover 'names' and then find appropriate objects for them. Language and world are disclosed together.[44]

Language is not an empty form but world-filled because it is the medium in which the world appears. So if the child learns to speak englishly she is inducted into the English language-world in which she will live and feel at home. We would usually just say she inhabits the English world, because language is the *medium* in which the world appears and consequently it becomes 'transparent' – 'We stare right through the signifier to what it signifies'.[45] Through language we are brought first into the public world of we-consciousness. Contrary to a widely held belief, we do not begin with a *separate* I-consciousness.[46]

Now that we have identified the concrete phenomenon as the language-world, we can see that it must take the dynamic form of 'the one and the many' which we found when we considered the diversity of language. But we must be clear that the 'language-world' refers to the shared world of intelligibility that is our cultural understanding of the world and ourselves – i.e. the hermeneutic world – and does not refer to *all* of human experience.[47] Language is the medium of a shared consciousness, the vehicle for a way of seeing which *is* a mode of appearance of the world. Differences in language will therefore disclose differences in the mode in which the world appears. These will be self-differences of the world and not different worlds – the world can be intensively multiple without becoming extensively many.

Understanding the difference between language as disclosure and language as representation is crucial here. If we say that language represents the world, we are conceiving language as being separate from the world. If we assume that we can have direct access to the 'world itself' independently of language, then language becomes an entirely secondary matter of just attaching labels to things. In this case different languages would simply be different ways of labelling one and the same world. But if, less naively, we assume that we don't have direct access to the 'world itself', but only to the world as seen through language, then each language-view is a worldview. In this case we can only have different perspectives of what we suppose is one and the same world as it is mediated to us through language. We cannot have direct, i.e. unmediated, access to the 'world itself', and as a consequence the language through which we see the world becomes at the same time something that comes between the world and ourselves. Here it is as if language is a window through which we see the world, but a window that has a structure of its own which it imposes on the world, so that we can only see the world as it is 'frameworked' by language. As in the naïve case, there is only one world, but now it is inaccessible in itself, so that all we can have are different language-views through which the world is mediated. In this case, language stands between us and the world, which we can therefore never know directly as it is 'in itself'. What both of these approaches share is the assumption that language is separate from the world, which means they conceive the relationship between language and the world in the framework of the subject-object separation. This is because they begin 'downstream' with the world conceived as a finished entity, instead of 'upstream' with the coming-into-being of the world which is its appearance in the disclosure of language.

In the disclosive mode of language, unlike the representational mode, there is no possibility of that dualism of world and language which conceives the world as separate from language, and yet at the same time considers language as a framework through which we must see the world. On the contrary, in the disclosive mode of language 'the world itself presents itself', so that 'what the world is not different from the views in which it presents itself'.[48] It is sometimes said that language is a prison from which we cannot escape to encounter the 'world in itself'. This is how it can seem when we consider language only in its representational mode. But the experience of language was the very opposite of a prison for Helen Keller – it liberated her from darkness and she 'saw everything with the strange new light that had come to me'. Far from closing her off from the world, the dawning of language opened her to the appear*ance* of the world.

Remembering that the concrete phenomenon is the language-world, we can see that the diversity of languages must entail the diversity of the world. This is quite different from the diversity of perspectives we have of the world as seen through the framework of different languages – which is how it seems to the representational mode of language, and which inevitably leads to relativism. In the disclosive mode, we do not see different perspectives of the world, with the danger that these will come to be thought of subjectively, as if they were many different worldviews. Since, in the disclosive mode, it is the world itself that presents itself, and not a representation, it would be better to say that it is the world itself that appears (read verbally) differently in different languages. In other words, what we have is not fragmentation into many different worlds, but the world *itself* manifesting differently according to the differences in language. So different 'worldviews' are not different views of the world, but different appearances (read verbally) of the world itself. The differences are self-differences of the world. Instead of many worlds, or many perspectives of one world, what we have is the dynamic unity of self-different modes of world. We have gone from the extensive to the intensive way of seeing, where once again we discover that what we see as diversity is in fact the unity. But to find this we have to think dynamically, and when we do we discover that what we mistook for relativism is in fact dynamics of being.

The significance of this may not be clear immediately if we limit ourselves to languages that are closely similar – for instance, English, French, and German – although even here there are more significant differences than at first we may realise.[49] But it becomes much clearer when we consider languages which are very different from those to which we are accustomed. This is why the attention of linguists has become focused increasingly on the languages of the indigenous people of the world. It is here that we find the greatest diversity of language, but also the greatest threat of extinction – it has been estimated that by the end of this century half of the present languages of the world will be extinct, which means the loss of a language every ten days. It is now recognised that when a language disappears a culture is extinguished: 'When a language dies, a whole world dies. It takes millennia to develop, and is an artefact that contains within it a whole culture. This is a tragedy'.[50] In recent years indigenous peoples themselves have spoken about the effect which the loss of their language has on them. Language is no secondary matter for them – as it might be if it were just a matter of representation. Often obliged to give up their own language in favour of another language which is alien to them, they have made it clear that doing so is not just a matter of switching to a different view of the same world: 'The worlds in which different cultures live are distinct worlds, not merely the same world with different labels attached'.[51] David Peat describes encounters with Native American people who emphasise that language is the key to their culture: 'Language, so traditional Indigenous people say, is the door to their world'. They tell him that 'the language of a people is their life', and that 'a people can no more live without its language than a tree can grow without its roots'.[52] When a language disappears a whole mode of world is lost and the world as a whole is diminished. It is remarkable how the understanding of the difference between disclosure and representation, and the relationship between language and world, that emerged in European philosophy during the last century, illuminates and is illuminated by the existential loss of language and world being experienced everywhere by indigenous people.

The Fallacy of the Proposition

But what about the idea of a universal language? What we have just seen *is* the universal language – i.e. the diversity of languages *is* the dynamics of the universal language. It is the concrete universal in which the whole comes to presence in the part, so that the part is an expression of the whole. But when we ask about the idea of a universal language we are usually not thinking of the concrete universal, which is more organic, but of the possibility of having one single language. Wouldn't this make it much easier for us all to understand one another? Clearly it would in one way; but it would quite literally be 'one way' because it would reduce the world to only a single mode. This may indeed be functionally appropriate in certain kinds of situation. But since it reduces the possibilities of the world, eventually a movement towards diversity will emerge again in order to compensate for the unavoidable one-sidedness of such a universal language. Latin fulfilled this role in Europe in the Middle Ages, which was followed by the diversification of Latin into the Romance languages (Italian, French, Spanish, Portuguese). Then in the eighteenth century it was French that became the universal language in Europe, which was followed in turn by a renewed emphasis on the diversity of languages, which we find exemplified in Herder's philosophy of culture, and which later came to expression in the rise of Nationalism in the nineteenth century. Today the role of universal language is taken by English, but by now we are less naïve about language and are therefore more open to the counterbalancing claim of the need for diversity. We recognise that any advantage which such a 'universal' language gives in the world of international business and finance, and other global institutions, is nevertheless always offset by an inevitable reduction in the possibility of meaning.

There is another way of trying to come to a universal language, and that is to construct a language artificially for this very purpose. The most direct way of doing this would be to construct the new 'universal' language out of the same forms that are found in existing natural languages. Esperanto is such a language, devised at the end of the nineteenth century based on roots common to the main European languages. Once again, although such a universal language, if adopted, would facilitate the communication of information and instructions, it would at the same time clearly lead us towards the reduction of the world to a single mode, and so to the impoverishment of the world instead of its enhancement. A more radical proposal for a universal language was put forward by the mathematician and philosopher, Leibniz, in the later part of the seventeenth century. His concern was the practical one of overcoming the differences between people which had led to the disaster and misery of the Thirty Years War (1618–48). He believed that the differences would disappear if a universal language could be constructed with meanings based directly on experience. Such a *characteristica universalis* (universal system of characters), as he called it, would allow us to express thoughts 'as definitely and exactly as arithmetic expresses numbers or geometrical analysis expresses lines'.[53] Yes, but how is this to be done? It seems that constructing such a universal language would require the agreement between people that such a language was itself supposed to produce, i.e. in order to construct a universal language there would already have to be a universal language. Since this is a self-contradictory situation, the only other possibility would be to assign meanings to signs by conventional agreement between those constructing the language. But this could only be done by means of existing natural languages. Hence this approach assumes that different groups of people, with different languages and cultural backgrounds, will nevertheless be sufficiently similar to converge to the same universal language – which means that, as before, the agreement which the process is supposed to produce must be there already at the outset. But since we don't have the proposed universal language *before* it is produced, how can we know that the different groups of people contributing *are* in fact sufficiently similar? As Toulmin puts it:

> Without independent assurance that different peoples
> perceive and interpret their experiences in sufficiently similar
> ways – as Leibniz said, that they 'have the same thoughts' –
> there is no agreement about the 'meanings' of the terms in
> our artificial language: without such prior agreement, there is
> no subsequent guarantee of mutual intelligibility.[54]

He concludes that 'the project of constructing a universal language is not difficult, as Leibniz concedes: it is downright impossible'.

Although it is now recognised that the attempt to come to the unambiguous expression of meaning by constructing an artificial language for that very purpose is impossible, the general aim of making language less ambiguous, and therefore more precise, is nevertheless one which occupied many philosophers in the first half of the last century. There seemed to be a fairly widespread view that natural languages are imperfect means of communication because they are far from being precise. It was felt that in science and philosophy, in particular, but also in politics and social affairs, the ambiguity which is part of natural language leads, not only to misunderstanding between people, but also encourages us to hold ideas which are really no more than 'illusions' of language. If this could not be improved by constructing an artificial language *ab initio* from which these confusions would be absent, the next best thing would be to try to improve existing natural languages by making them more precise and therefore more suitable for logical reasoning. Although the movement in this direction began, in central Europe especially, in the later part of the nineteenth century, it seemed to reach to a climax in the 1920s with the self-proclaimed 'Vienna Circle'.[55] The aim of this group – the influence of which spread far beyond its own immediate time and place – was to undertake a thorough logical analysis of the language in which the empirical sciences (especially physics) are expressed. They believed that this would enable them to formulate clearly and precisely what they called 'the scientific world-conception' (i.e. not just individual sciences as such, but science as a way of conceiving the world as a whole), as well as providing the means to criticise, and ultimately reject, the claims of metaphysics as being no more than consequences of the ambiguities inherent in ordinary language. Central to this programme for the reform of language is *the primacy of the proposition.* This is so important to the difference between the logical and the hermeneutical approaches to the philosophy of language, that we must go into it in more detail.

A proposition is what is said by a predicative statement of the form 'S is P', where S is the subject and P is the predicate which asserts something about, or attributes something to, the subject. For example, 'the cat (S) is on the mat (P)'. If it is true that the cat is on the mat, then this proposition is true and we say that it is a fact. Sentences asserting propositions are certainly by no means the only kind of sentence in ordinary language (for instance, 'I promise to come' is not a proposition) – and indeed for the purposes of everyday life they may not be very useful at all – but they are important when the focus is on science, because empirical science is concerned with facts. This is why so much attention has been given to the logic of propositions; it is clearly of central importance if we are concerned with 'the scientific world conception'.[56] Furthermore, the development of symbolic logic from the middle of the nineteenth century onwards, meant that propositions could be represented in symbolic notation, as in algebra, so that the correctness, or otherwise, of logical reasoning could be verified from the symbolic form alone regardless of the specific content. This 'mathematical logic', as it came to be called, was of particular interest to mathematicians and scientists, and its further development by Russell and Whitehead, and others, was readily adopted by the Vienna Circle as a key component of their programme for the logical reform of language.[57]

Emphasis on the proposition as the primary form of language, which became such a central feature of analytical philosophy in the twentieth century, goes back to Aristotle's discovery of logic as the science of the forms of valid reasoning in the fourth century BC.

The aim of logic – or 'analytics' as he called it – is to devise a system of formal inferences in which any valid inference could be expressed. If this can be done, then we can know whether an inference is valid or not simply from its form alone, i.e. independently of the subject matter which is the specific content of the inference. For example:

> All bachelors are unmarried men
> All unmarried men are mortal
> All bachelors are mortal

This does not seem particularly interesting, until we notice that we know that it is a valid inference without even knowing what 'bachelors', 'unmarried men', or 'mortal' mean, and *this* is what interested Aristotle. He was justly proud of this achievement, and his work on logic had a great influence throughout the Middle Ages in Scholastic philosophy, down to the later part of the nineteenth century when it began to be replaced by modern symbolic logic. Because Aristotle proposed that all knowledge could be expressed in the form of a subject-predicate sentence, the propositional form, 'S is P', became fundamental in the western philosophical tradition.

This discovery did not just fall from the sky into Aristotle's lap. In fact he first came to his system of logic in a way that is usually unsuspected today, when we are accustomed to thinking of logic as embodying *universal* principles of human reason. He discovered it by observing the thinking of the mathematicians at the Platonic Academy in Athens. Kline expresses it succinctly:

> Aristotle abstracted the principles of deductive logic from
> the reasoning already practised by the mathematicians.
> Deductive logic is, in effect, the child of mathematics.[58]

This is astonishing because it means that this logic, far from being truly universal, is founded on the particular practices of the mathematicians in a highly influential school of mathematics.[59] The remarkable step taken in the Platonic Academy – and which Aristotle will have experienced at first hand – was the discovery of mathematical proof by deductive reasoning. This makes possible an ontology of mathematics, which distinguishes mathematics from any kind of empirical activity, and as a consequence for the first time enabled mathematicians to understand their activity in its own terms, so as 'to be able to say what they are dealing with, and to make evident that what they were doing was in any case not some sort of physics'.[60] This important point is still often not grasped, especially by those who do not themselves have a mathematical background. Yet it is crucial:

> It is important to appreciate *how radical the insistence on deductive proof was.* Suppose a scientist should measure the sum of the angles of a hundred different triangles in different locations and of different size and shape and find that sum to be 180° to within the limits of experimental accuracy. Surely he would conclude that the sum of the angles of any triangle is 180°. But his proof would be inductive, not deductive, and would therefore not be mathematically acceptable.[61]

A proof which would be mathematically acceptable would be one that did not involve measurement at all. It would be given entirely in terms of relationships between the angles without any need to refer to the actual size of the angles in a particular triangle. Consider any triangle ABC with angles *a, b,* and *c* (see Figure 14).

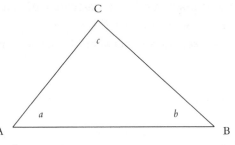

Figure 14. The triangle.

Extend the side BC into a straight line, and draw a line through vertex A parallel to this line (see Figure 15).

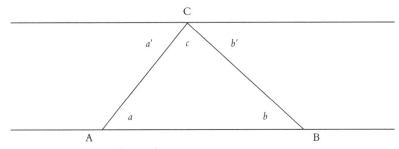

Figure 15. Constructing the proof.

Angle *a* equals angle *a'* because they are alternate angles between parallel lines. Angle *b* equals angle *b'* for the same reason. But angles *a'*, *b'*, and *c* must add up to 180° because they make a straight line. Hence it follows that angles *a*, *b*, and *c* must add up to 180°. In such a deductive proof we see that the angles of a triangle *must* add up to 180°. This is entirely different from just saying that the angles of a triangle do in fact add up to 180°. It's not that they happen to do so – as if this were an empirical discovery – but that they cannot not do so.

This is the difference between the mathematical and the empirical that Plato was so concerned to establish, and which his contemporaries found so hard to grasp:

> Plutarch relates ... that Eudoxus and Archytas, famous contemporaries of Plato, used physical arguments to 'prove' mathematical results. But Plato indignantly denounced such proofs as a corruption of geometry; they utilized sensuous facts in place of pure reasoning.[62]

Gadamer believes that it was the creative advancement of mathematics at that time and place, by those particular people, that provided the motive for Plato to introduce the so-called *chorismos*, which is often conceived as a 'separation' between the sensory and the ideal. As Wachterhauser says:

> Such a distinction is particularly essential to mathematics if mathematicians are to understand their own thought by its own inherent standards and not mistake it for a type of empirical research, to which apparently even a gifted mathematician like Theatetus was susceptible.[63]

So when we are doing mathematics we must make a clear and sharp distinction so that we do not confuse the noetic aspect of things with their sensory aspect. But this does not mean that these aspects are separated ontologically into two different worlds: a sensory world of appearances and an intelligible world of ideas. When we understand the 'separation' in the context of mathematics, we can begin to see how what is really a necessary methodological distinction has been turned into an unwarranted ontological bifurcation. As Plato says in the *Parmenides* (142 d-e), if there were a second world of ideas separate from the world of appearances, then it would be a consequence of such an ontological divide that this other world of ideas would exist only for the gods, and the sensory world of appearances only for us. Gadamer emphasises that the ideas are ideas *of* appearances, and consequently that there is a fundamental connection between appearances and ideas which enables us 'to speak of them as two sides of the same reality'.[64] In which case it is a mistake to attribute a two-world ontology to Plato – which of course is just what always has been attributed to Plato. The noetic side of things is just as much an aspect of the one world as the sensory side – it's just that they need to be 'separated' so that we do not confuse mathematics itself with some kind of empirical investigation. As Gadamer says: 'One needs only keep in mind what Plato had in view and the historical motivation that led him to carry out this separation of the ideas from the appearances'.[65]

One of the formal procedures of deductive proof that the Greeks developed argues that a proposition must be true if the negation of the proposition leads to self-contradiction. In this case the negation must be false, from which it is concluded that the original proposition must be true. An illustration of this procedure may be useful, because it is from this kind of proof that Aristotle first abstracted the principles of logic from the thinking of mathematicians. We can take the example of the proof that there is no ratio of two integers p/q which when multiplied by itself will give 2. Such a ratio of whole numbers p/q is called a rational (ratio-nal) number, and so what is asserted here is that the square root of 2 (i.e. the number which when multiplied by itself gives 2) is not a rational number. It is therefore said to be an irrational number. We can easily see how this number arises by considering one of the simplest geometrical figures, namely a right-angled triangle with equal sides each having unit length. What is the length of the hypotenuse? From the theorem of Pythagoras we know that the square of the length of the hypotenuse must be 2 units. In which case the length must be the square root of 2. Now what is unexpected is that if we try to express this length as a rational number p/q, we find that we just cannot do it. But what is remarkable is that we do not have to flounder around empirically trying to find two whole numbers, p and q, whose ratio when multiplied by itself will give 2, because we can prove mathematically that this *cannot* be done. The procedure is to begin by proposing that it *can* be done – i.e. that the proposition that the square root of 2 can be expressed as a rational fraction p/q is true – and then to show that this leads to self-contradiction and so must be false. The procedure is then to say that if a proposition is false the negation of that proposition must be true, so that if it is false that the square root of 2 can be expressed in the form p/q, then it must be true that it cannot be so expressed. This completes the mathematical proof, and just as a proof in geometry requires no measurements to be done, so this proof is done without any calculations – because it is mathematical and not empirical.[66]

This is the kind of thinking from which Aristotle first abstracted the principles of what came to be called logic. In the first place this was therefore an abstract formalisation of Greek mathematical thinking, but which later became 'laws of thought' and were hypostatised into 'universal principles of reason'. Traditionally three fundamental principles were identified:

(i)   The principle of identity: P is P.
(ii) The principle of non-contradiction: not at the same time P and not-P.
(iii) The principle of excluded middle: either P or not-P.

Although these are stated as three separate principles, they are really more like three different aspects of one principle. No matter which one we choose, we soon realise that it entails the other two, so that they are not really independent propositions. The first one, P is P, seems to be so trivially true that at first we might even wonder why anyone would mention it in the first place. Yet, as we shall see, it says much more than we might think, and is in some ways the most important. From what has been said about the procedure of proof, we can easily recognise both 'not at the same time P and not-P' and 'either P or not- P' in the form of thinking. It is these principles, grounded in the procedure of proof in Greek mathematics, which were for so long taken to be universals of reason and as such not open to question.[67]

Yet there is something much too sharply defined about this 'logical' thinking. In the first chapter, when we tried to catch distinction 'in the act' we found that this is a unitary act which {differences/relates}. If something is distinguished – call it A – it is thereby *internally* related to what is not-A in the very act by which it is distinguished in the first place. The act of distinction which differences simultaneously relates. Thus what A *is* necessarily entails what A *is not*, so that in being what

160

it is, A is not independent of what it is not. Since we *cannot* have A without at the very same time having not-A, if we try to make A entirely self-identical, the distinction would simply disappear because there cannot be difference without relation. We can see from this that there is no such thing as an independent proposition, i.e. in which the meaning is completely contained in the proposition, because such a proposition would be entirely self-identical and therefore related only to itself. Such a self-contained unit of meaning could not even be asserted in the first place, because to assert a proposition is to make a distinction which {differences/relates}, and thereby necessarily entails a relationship to what is not included specifically within it. What is said is necessarily related to what is not said by the very act of saying. The paradox is that the only way we could have a 'pure' proposition, which contained its meaning completely within itself, would be to have only what is said without what is not said, and we could only have this by saying nothing at all. In other words, there can be no such thing as a self-identical proposition which is complete in itself, containing its own meaning without any reference beyond itself. But isn't this just what the principle of identity asserts that there can be? At first the assertion that 'P is P' may seem to be merely tautological, and therefore not to say anything. We may even wonder why it would ever be mentioned in the first place, let alone elevated to the status of a principle. But far from being trivial, no more than an empty tautology, the principle of identity asserts something fundamental, which is that a proposition contains its own meaning and therefore can be understood completely in terms of itself. It is this principle which really forms the basis for the other two principles, and which taken together are held to be *evident* because 'thought could not deny them without denying itself'. But 'the one whose authority is the most assured is the principle of identity, which in fact has hardly ever been disputed'.[68] On the contrary, this is precisely what the hermeneutical experience of language calls into question.

Mathematics, logic, and language are strangely entangled in Western philosophical thinking. Logic, the child of mathematics, is taken as the model for understanding language. This leads to the primacy of the proposition conceived as an independent, self-contained unit of meaning. But the moment we step outside of the mathematical-logical conception into the lived experience of language, we discover that the hermeneutical experience of understanding cannot be reduced to the meaning of such a propositional statement. Gadamer emphasises that 'there is no proposition that can be comprehended solely from the content it presents'.[69] What is said does not encapsulate its own meaning, as if it could be fully understood independently of the context in which it is said – where 'context' refers to everything that is meant 'with' the text (con-text) but which remains unspoken.[70] What is said 'carries with it the unsaid', i.e. what is not said but is intended along with what is said. This is not some deficiency of expression, but what Gadamer calls 'the living virtuality of speech', which he says 'brings a totality of meaning into play, without being able to express it totally'.[71] What is not expressed is just as much part of the meaning as what is expressed – indeed it may even be the more important part, because what is expressed can only really be understood in the context of what is not expressed. The meaning is what is intended – in French it is the *vouloir dire,* the 'want to say' – but which is always more than can be said, even though we try to say what we mean. Anyone who speaks or writes knows that their words do not express the whole of what is meant. What is intended, the meaning, can never be expressed completely as the content of a statement, i.e. as a proposition, because the context of what is said is just as much part of the meaning, and evidently this cannot be included in the content. Yet without the context there would be no possibility of saying what is said in the first place. The idea of the proposition as a self-contained unit of meaning (A is A) is therefore an abstraction. What we find instead of the primacy of the proposition is the lived experience of language – hermeneutics instead of logic. We can now understand why, for Heidegger and Gadamer, the proposition is a form that is secondary and derivative, and not primary as in logic. Gadamer describes the 'construction of logic on the proposition' as 'one of the most fatal decisions of Western Culture'.[72]

This certainly should not tempt us into thinking that we are excused from trying to say what we mean as clearly as we can. But it is a sheer fact of experience that what is meant is *always* more than we can say. We try to say what we mean, but even on those occasions when we do feel that we have done, we soon begin to realise that somehow it seems to fall short of what is meant. It isn't that we have just missed something out, and if we could include this we would be able to say completely what is meant. Meaning is not additive like this. The meaning is the whole which comes to expression through what is said – we could say that it presences in what is said, but not that it becomes present. But this whole is not the totality of what could be said, as if we could just add on more and more until we reached all that could be said. The whole is not the same as the totality – which is only the extensive notion of the sum total – because the whole comes to presence within the part without becoming present as such, i.e. without becoming an object which 'stands out' and is therefore purely present. So the part is an expression of the whole – and not just a component of the whole as it would be extensively – so that the whole always comes to expression part-ially and never completely, even though it *is* the whole that comes to expression. Such a part that is an expression of the whole is a 'whole-part'. This is possible intensively but not extensively – otherwise it would only be part of the whole, instead of the whole manifesting partially. If the whole manifests partially in this manner, then the part itself is an expression of the whole (a whole-part and not part of a whole). Thus, in the case of the meaning that comes to presence in what is said, this can only be expressed partially through what is said. Each expression of meaning is complete (it is a whole-part) but the expression is always necessarily unfinished. If this were not so, the whole would be the same as the totality and the meaning would be finished. The result would be a static condition in which meaning ceased to be a dynamic whole and became a fixed entity. But this is just the condition of the proposition, which is therefore once again seen to be an abstraction from the hermeneutical experience of language. What is primary is not the self-contained static proposition of logic (A is A), but the dynamic whole-part of hermeneutics.

It is clear from this that when we try to understand a written work, it will not do just to focus on what is said in the content of the statements we read. As Heidegger says, we must develop a hermeneutical understanding of language: 'what is essential in all philosophical discourse is not found in the specific propositions of which it is composed but in that which, although unstated as such, is made evident through these propositions'.[73] We only understand what is said when we recognise the whole (the unsaid) manifesting through the part (the said), so that the part is understood as an expression of the whole – i.e. as a whole-part. Hence again we see that the hermeneutical whole-part is primary, and that the proposition can only be secondary because it presupposes the hermeneutical dimension of language, which is the background of the unsaid from which what is said comes-into-being as the part that expresses but cannot contain the whole.

163

6. Taking Appearance Seriously

The mathematician and philosopher, Alfred North Whitehead, famously said:

> The safest general characterization of the whole Western philosophic tradition is that it consists of a series of footnotes to Plato.[1]

Although this is clearly an exaggeration, it is surprising how perspicacious it is when taken in a sufficiently broad manner. It becomes all the more pertinent when we take into account the fact that the influence of 'Platonism' extends far beyond philosophy.[2] Augustine, in many ways the founder of western Christianity, introduced Platonism into Christianity to such an extent that Nietzsche described it as Platonism for the people.[3] Furthermore, there is the influence that Neo-Platonism had on the transformation to the heliocentric planetary system, and the development of mathematical physics in the seventeenth century, which we looked at briefly in the second chapter. This was a major influence behind the idea that there are mathematical 'laws of nature' which organise matter, and thus determine the structure of the physical world, but which are themselves separate from and superior to the matter they act upon.[4] In the context of the time, such *transcendent* mathematical laws were easily conceived as being 'the thoughts of God', so that mathematical physics could be presented as a quasi-religious activity which 'recast God as a mathematical creator'.[5]

The Platonic tradition in Western philosophy introduces a fundamental dualism that separates being from appearance. Wherever it occurs, Platonism is characterised by 'an opposition between the multifarious appearances involved in perpetual change and an immutable realm of existence, forever persisting in strictest self-identity'.[6] Knowledge of 'Being-as-it-is-in-itself' (*ontos on*) is considered to be the only true knowledge, whereas everything else is only belief and opinion about changing appearances. The characteristic of genuine knowledge is that:

> Since it is concerned with Being-as-it-is-in-itself, it is free from all relativity with regard to subjects, their standpoints, and the vicissitudes of their lives. Because of the persistent self-identity of this Being, genuine knowledge is perpetually true, under all circumstances and for everyone.[7]

The key thing for Platonism is that *difference is excluded from being* – and consequently also from genuine knowledge. Difference belongs only to the realm of appearance – which is separate from being. Because there is no 'difference' in being – it is always self-identical, the self-same – there cannot be any movement in being, which is therefore static. Here we have the standard picture of metaphysics, in which there is the separation of being from all that is dynamic, so that everything falls apart into a realm of being without change and a realm of change without being – which is a realm of 'mere appearance'. But if, contrary to this metaphysical picture, there can be difference within being, then being will become dynamic. This brings us to the notion of self-difference, and the dynamic unity of self-differencing instead of the static unity of self-sameness. But self-difference cannot be extensive difference, because this would be the difference of one thing from another, whereas self-difference means that something becomes different from itself while remaining itself. Self-difference is an *intensive* distinction, which means that it is a difference within One which does not result in two and yet is not the same as one. Self-difference brings us into the intensive dimension of One, instead of the extensive dimension of many ones, where we have to learn to think in a way that does not separate into two and yet does not reduce to one.

Unfamiliar though this dynamic understanding of being may be, it is the ontology that emerged in phenomenology and hermeneutics (and also in the work of Gilles Deleuze) in European philosophy during the twentieth century, and which has produced a sea change in western thinking that has not always been sufficiently recognised. It is the ontology that emerges when we leave the end result – the final stage – behind and instead move attention 'back upstream' into the dynamic phase of coming-into-being. Phenomenology is a shift of attention within experience, which draws attention back from *what* is experienced into the *experiencing* of what is experienced. We saw in the first chapter that this is more subtle than it sounds at first. By focusing on the act of distinction – which means drawing attention back from *what* is distinguished into the *distinguishing* of what is distinguished – we discovered that the *happening* of distinction is the appearing of what is distinguished. This takes us to the heart of phenomenology: the phenomenon is not only something which appears, but which appears *as appearing*. So the phenomenon is not merely the appearance as we usually think – but the appear*ance*. Once we recognise this, we get a sense of the phenomenon as something 'coming into being' – not in the metaphysical sense of coming into existence, but in the phenomenological sense of being as the appearance of what-is. When Heidegger says 'being means appearing' – and adds that 'appearing is not something subsequent that sometimes happens to being' – he could equally well have said 'appearing means being'.[8] The dynamic approach brings us to the point of recognising, albeit with some astonishment, that appear*ance is* being – which is certainly not true for appearance. The endless discussion in philosophy about being and appearance gets nowhere, because it focuses on appearance, which is too late, instead of shifting back 'upstream' into appear*ance*. When we take this step, we find that the metaphysical separation between being and appearance disappears, but without reducing being to appearance.

By discovering the dynamic depth in the appearance which is the appear*ance*, we realise that it is the thing itself manifesting directly in its appear*ance*, and not simply a representation of it 'in consciousness' in the Cartesian sense. When we shift 'upstream' from appearance to appear*ance*, we leave behind the subject-object separation, and all the problems of epistemology that come with this, and enter into a non-dual condition in which what manifests *is* what is, without any mediation by intervening entities of some kind (images, representations) in consciousness. We have seen this in some detail in the case of understanding the meaning of a text. In the fourth chapter we saw that there is no separation between meaning and understanding in the event of understanding, because understanding *is* the appear*ance* of meaning. We also found that, as well as the phenomenological description, Aristotle's single-event description of understanding meaning leads us to the same non-dual condition. When we shift attention from appearance to appear*ance,* we realise that there are differences in understanding which are not merely subjective (some will be, of course) but are manifestations of differences in the meaning of the work itself. These are therefore the work's self-differences, and as such they belong to the work as its own possibilities – though *not* in a preformed way (to avoid this confusion, it might be better to follow Deleuze and speak of the work's 'virtuality of meaning' instead of referring to 'possibilities'). The dynamic form of the 'one and the many' that we explored in Goethe's way of seeing the living plant, now turns out to be the form of the 'one and the many' in hermeneutics – where the differences in meaning are now self-differences in appear*ance.* So the diversity of interpretations *is* the dynamic unity of the work. Such an 'organic' hermeneutics is clearly free from the dichotomy of objectivism (a single meaning) and relativism (a plurality of meanings).

When we turn to language itself, the shift of attention 'upstream' from *what* appears to the appear*ance* of what appears, takes us into the difference between language as disclosure and language as representation. It is very clear here that, although we may talk about language as a prison which shuts us off from the world, this is true only in the case of the representational mode of language. In the disclosive mode, on the other hand, language opens us to the appear*ance* of the world. We explored this in some detail in the previous chapter, where we saw that different 'language-worlds' are not different representations of a 'world in itself' which is independent of language, but self-differences in the appear*ance* of the world, and hence different modes of world. They are self-presentations of the world in which the world itself presents itself. So instead of a pre-given 'world in itself' that is represented in language – which would therefore be an intermediary factor between the world and ourselves – we now see that language is the medium of the self-manifesting of the world. This is the reversal we find when we shift 'upstream' from language as representation to language as disclosure. Language is no longer an obstacle between ourselves and the world, but the very means by which the world itself comes to appearance. But the world is no longer single in the disclosive mode – as it is when the world is conceived in the 'downstream' manner as an object – but manifests self-differently in accordance with the language. Instead of different perspectives of a single world, we have the world appearing as self-different modes of itself. As such the world is intensively multiple, without falling apart extensively into 'many worlds'. Here also we can see the similarity to organic life, and the way that in the 'logic of life' self-difference replaces self-identity in the 'logic of bodies'.

This event of coming-to-be in language is different from the ideal of objective knowledge in science. For this reason it has often been dismissed as simply being 'subjective'. Truth is what is discovered by science, it is thought, and as such it takes the form of being the very same for everyone. What science achieves could be called the 'view from everywhere', which is the 'view from anywhere' – not the 'view from nowhere', as is sometimes misleadingly said. We can see this very clearly in the universalism of the mathematical style of thinking which has gradually dominated since the time of Descartes – and which is now applied so widely that we just take it for granted, even though there are many kinds of situation where it is highly inappropriate. What we can call the 'hermeneutic style of thinking' turns this inside out. What looks like the sheer plurality of many different viewpoints, and hence seemingly subjective, becomes instead objective manifestations of something coming-to-be differently in different contexts and situations. Instead of the abstract universal of the mathematical style, we have the concrete universal that is more 'organic', where, what otherwise seems to be just a plurality, is actually the dynamic unity of self-differencing. There is a universality in the dynamic approach that is characteristic of hermeneutic thinking, but it is not the abstract universal we are always looking for, and so it easily gets overlooked and mistaken for 'relativism'.

There now seems to be a fair measure of agreement among historians and philosophers, that the beginning of modernity is to be found in the seventeenth century with the revolutionary step of basing the natural sciences on mathematics. The development of the new science of mathematical physics, from Galileo to Newton, introduced a new way of understanding nature which was soon seen to be astonishingly successful. A key factor in this development was the notion that there are universal laws of nature which apply in the very same way to everything – there is no place in such laws for the differences between things. The kind of universality that is characteristic of mathematics and the mathematical laws of nature, also began to seem attractive for other reasons in the seventeenth century. This was a time of murderous religious disputes, culminating in the savagery of the Thirty Years War between Protestant and

Catholic powers. The devastation of this war was felt over most of Europe, but especially in Germany where every third person was killed over differences of religion. Toulmin points out that it was against this background that the kind of agreement and certainty found in mathematics began to have an appeal beyond the confines of mathematics itself.[9] Euclid's *Elements of Geometry* had been translated into different languages, and had become widely available as a result of developments in printing – for some time it was the most widely bought book after the Bible. This provided a model for rational thinking. For example, the method of *proof* enables us to be *certain*, and therefore to *agree*, that the sum of the internal angles of any triangle *must* be equal to two right angles (180 degrees). This does not depend on drawing triangles and making measurements (room for disagreement here), but follows of necessity from the very idea of a triangle – i.e. it is intrinsic to the concept 'triangle'. It is discovered by reason, not by measurement, and this is why it is certain and consequently not subject to disagreement. Here then is an example of a situation where everyone will think the same – Protestant triangles are not different from Catholic triangles. The idea began to develop that, if the mathematical *style* of thinking (not necessarily mathematics) could be extended to other areas outside of mathematics itself, then it would also be possible to reach certainty, and hence agreement, in those areas by the same kind of reasoning that is used in mathematical thinking.[10]

But it was the impressive success of mathematics in the new astronomy and physics which was to have the biggest impact – especially after Newton.[11] What captured the imagination was the idea that there are mathematical 'laws of nature', which are universal in the sense that they discover a unity in the diversity of phenomena from which all differences are excluded. For example, Newton's universal law of gravity applies in the same way to *all* bodies, regardless of differences in material constitution. The idea began to develop that, as well as nature, there is also human *nature* – i.e. the idea that human beings are 'of a piece with nature'.[12] Hence there should be a science of human nature which would apply to all people regardless of their evident differences, which would enable the solution of social, political, and moral problems in a rational way that all could agree on – as they were agreed about mathematics and physics. This notion that there should be universals in human nature – the very same for everyone – offered the utopian hope that it would be possible to build a rationally organised society in which everyone would be in agreement. But of course, this dream of universal reason that looked so attractive to some, particularly in view of the social context of the time, must inevitably lead, not to the promised land of what is universal, but into the cul-de-sac of uniformity because it is based on the exclusion of difference.

Historically this resulted in the division in western culture between the world of Science, with its emphasis on the mathematical, and the world of Language. The metaphysical foundation of science underpins the notion that there is a single Truth underlying the plurality of phenomenon, a single ultimate reality behind the appearances. We are so familiar with this idea now that we barely notice it.[13] This has the consequence that, whatever does not conform to this ideal of what is true knowledge, seems inferior and second class, to be redeemed only to the degree to which it can be made over in the image of science. But with language there comes difference, and this is inherent to language, not something that can be replaced by a common factor. Thus different languages are not just alternative ways of expressing the same thing. What something 'is' is not pre-given, but is 'expressed' in the dual sense of made manifest and given determinate form by the language in which it is disclosed. If language were only representational, then it would be the case that different languages would simply be different ways of representing the same thing – which would be pre-given, even though we may not have access to it directly but only as mediated by language. But language is primarily disclosive, which means that what appears is not pre-given and yet it is the appear*ance* of what is. We have to think dynamically if we are to get this right.

In the event of disclosure, the thing itself comes to be in language – which is not to be confused with the supposed metaphysical notion of the 'thing-in-itself' that is forever beyond appearance. This coming-to-be-in-language is the self-presentation of the thing itself:

> The ontological significance of language is that the thing
> itself presents itself ... as a perspective of itself.[14]

It does not appear absolutely in any particular language, but 'presents itself in every particular language as a perspective of itself' and 'this is the most complete expression of itself possible in that particular language'.[15] In other words, it presents itself differently in each particular language, but these differences belong to the self-presentation of the thing itself as perspectives of itself – not as pre-given, but dynamically as coming-into-being. However, we must be careful here not to fall into thinking of 'perspective' in the representational manner, i.e. as if the thing is given already and we are looking at it in different perspectives through different languages. Instead we must remember that it is the thing coming-into-being as itself in the mode which is made possible by each particular language. The differences are *self-differences* of the thing itself in the 'organic' sense – they are not preformed. In other words, language is ontological in the event of disclosure, and difference is *included* in the ontology of language – *difference becomes ontological,* which is impossible in metaphysics. This is unexpected because, in the metaphysical tradition, difference is strictly excluded from being and relegated to the secondary level of what consequently become 'mere appearances'. The ontology of science is metaphysics (what is real is hidden behind what appears), but we can now see that metaphysics is not the only possibility – which is what was assumed in the western tradition until Heidegger and Gadamer showed otherwise. Once difference becomes ontological, appearance is taken back into being, so that what appears is no longer conceived as hiding what is real behind it. The 'curse of mereness' is lifted and appearance becomes ontological in its own right.

This liberates us from a restrictive pattern of thinking in which we have been trapped for a long time.[16] It rebalances the one-sided cultural emphasis on science and mathematics as the *only* way to truth, and restores to language the ontological significance which had been lost sight of in the glare of metaphysics. Everything that has been said about language and the world applies equally to language and the written text. So this ontology of language restores to literary and philosophical works the possibility that they have their own truth. This is the truth of disclosure which draws things into appear*ance*. It is clear that our role in this is not that of a subject in front of an object, but that of a participant in an event of appear*ance*. For Heidegger, 'to be' means to appear or be manifest. But 'beings cannot be manifest without a clearing or opening in which their self-manifestation can occur'.[17] We, in our cultural-historical existence, are the openness for the appear*ance* of what is. But we must be careful here, because if we say that we are the openness or clearing where things appear as what they are, it is only too easy to introduce a dualism between the opening and what appears in it. We unwittingly introduce a separation, as if the opening is pre-given and what is appears in it. But the opening *is* the appearing. So it would be better to say, not that we are the place where what is appears, but that we *are* the appear*ance* of what appears. As Sheehan puts it, drawing on the language of Aristotle, human being (in our cultural-historical existence) is:

> Not just the *topos eidon* – the place where meaning appears –
> but above all the *eidos eidon*, the very appearing *of* appearance.[18]

This is where we come to if we take appearance seriously. In doing so we find that phenomenology retrieves an earlier stream of philosophy from before the modern period beginning with Descartes. According to the Cartesian-Lockean tradition, which still dominates our culture, what we are directly aware of is not the world, but only the representation of the world which exists in the form of ideas in our minds. In this case, what the mind knows is only its own ideas, and we are locked into our own subjective experience. For Aristotle, on the other hand, when we understand the meaning which something *is* (the what-it-is), then our understanding just *is* that meaning. So, in the event of understanding, what is understood becomes itself in us – as we have described in Chapter 4 – which is why Aristotle says that 'the soul, in a way, is everything'.[19] This is the same point to which phenomenology brings us in its own way. It also brings us to a different understanding of the self. We no longer understand ourselves as a self-centred subject facing the world of objects 'out there'. Instead we become a non-subject-centred self open to the world. We discover ourselves as 'datives of disclosure, as those to whom things appear'.[20] Phenomenology brings us to marvel at the 'fact that there is disclosure, that things do appear', and that we 'serve as datives for the manifestation of things'.[21] This is what Heidegger, following Husserl, calls 'the wonder of all wonders'.

In fact, we are more familiar with the difference between appear*ance* and appearance than we may think at first. For example, think about a gesture. Imagine we are walking down a street, and someone on the other side raises their arm. What do we see? Do we just see someone raising their arm, or do we see someone waving at us? It might be one or the other, but what is the difference? In one case we see just a physical movement, whereas in the other we see 'hello!' Instead of merely a physical movement, we *see* meaning – the gesture is an *expressive* movement, whereas the movement is just a movement. The difference is between appear*ance* and appearance. In the immediacy of the *lived experience*, we do not just encounter the appearance but the appear*ance*.

However, we don't usually think of it this way. Instead, we think that what we actually *see* is the appearance – the movement – and that we then *add* the meaning on to this to recognise this movement as the gesture 'hello!'. This is how it seems to us after (even a very short time after) the lived experience, when the immediate presentation has already become a re-presentation. This is when the lived experience which is mediated through the right brain, is passed to the representation of experience by the left brain. When this happens, what is originally whole is represented in terms of a separation that is not there in the lived experience. Referring to the primacy of wholeness, McGilchrist says:

> The right hemisphere deals with the world before separation,
> division, analysis has transformed it into something else,
> before the left hemisphere has *re*-presented it. It is not that
> the right hemisphere connects – because what it reveals was
> never separated; it does not synthesise – what was never
> broken down into parts; it does not integrate – what was
> never less than whole.[22]

If we don't get this the right way round, we cannot help but think of a gesture as a mental meaning *added* to a physical movement, whereas it is in fact *lived* meaning.

Similarly, we think that someone making a gesture first has an inner (mental) meaning which she then 'expresses' outwardly as a physical movement. But this way of thinking is already 'too late' for the lived experience. Once again, it thinks of a gesture as appearance plus meaning, as if they were separate and brought together. But the gesture *is* the appear*ance* of meaning, not meaning added to appearance. Gadamer says it very clearly:

> What a gesture expresses is 'there' in the gesture itself. A
> gesture is something wholly corporeal and wholly mental at
> one and the same time. The gesture reveals no inner meaning
> behind itself.[23]

A gesture of anger *is* the anger – it is the appear*ance* of anger. The anger is not behind the gesture – as if it had to be added to the shaken fist to make this a gesture. Once again we have to learn to think in a way that does not separate into two, and yet does not merely reduce to one. In this way we can walk the tightrope between dualism, on the one hand, and reductionist behaviourism on the other.

Wittgenstein sees all modes of behaviour in this way. He sees behaviour *as lived* as being *intrinsically* expressive, and not 'just' behaviour with the expression added on. This is what he says about emotions:

> 'We *see* emotion.' – As opposed to what? – we do not see
> facial contortions and make inferences from them (like
> a doctor framing a diagnosis) to joy, grief, boredom. We
> describe a face immediately as sad, radiant, bored, even when
> we are unable to give any other description of the features.[24]

The facial contortions – which we may *choose* to focus on – are an abstraction from the primacy of the whole expression. What we see is not just the physical appearance – which we then have to interpret – but the appear*ance* of the emotion. In lived experience (right brain) the emotion is there in the expression. But when the experience is re-presented (left brain) – and this will happen as soon as we think about it – we *separate* the emotion from the expression and think of it as being in an 'inner world' behind the physical appearance (the facial contortions). Now we are back in the unnecessary doubling of dualism, and the only way out of this seems to be to go to the opposite extreme of reductionist behaviourism. Either way, the lived experience of appear*ance* is lost. The result is that, on the one hand, we are plunged into the scepticism of the so-called problem of other minds (how can we really know that there are mental states behind the physical appearances?), whereas on the other hand we effectively reduce human beings to automata.

Wittgenstein developed a refined sensitivity to people's faces and voices: 'this kind of sensitivity can be gained only by experience – by attentive looking and listening to the people around us'.[25] It is this practice that is the basis for Wittgenstein's remark:

> Consciousness in another's face. Look into someone else's face,
> and see consciousness in it, and a particular *shade* of conscious-
> ness. You see on it, in it, joy, indifference, interest, excitement,
> torpor, and so on. The light in other people's faces.[26]

In such an experience we *see* a living conscious person. The *person* appears – we do not just see a physical body to which we attribute consciousness. In this experience, consciousness in another becomes visible – it's not in some 'inner world' behind the appearance, but *there* in the appear*ance*. This is when we see another person *as a person*, and not just as an animated being, or as if they were reduced to no more than an automaton – which it is only too easy to do in the conditions of the modern life. Of course, such sensitivity could be too much, and to get through the demands of daily life we often do need to 'switch off', and revert to a more pragmatic approach in which behaviour is not experienced as being expressive in this way. This may be how we get by, but we should not base our understanding of what it is to encounter another human being on this restricted condition.

Finally, there is the question of whether this approach could be extended from human being to natural being? Could we talk about the appear*ance* of nature, and mean by that the experience of the 'livingness' of nature? In other words, could we encounter nature as expressive being instead of an object? It has been suggested that Wittgenstein's approach could also be practised with nature to bring us to experience 'the being of things making itself manifest to us'.[27] One way this might be done is through Goethe's practice of active looking into nature followed by exact sensorial imagination – which certainly seems to be in tune with Wittgenstein's 'attentive looking' and 'observant sensitivity'.[28] One young Goethean researcher reports her experience of practising this way of seeing with the Nettle:

> After having spent time observing various nettles, going
> to and from them, eventually I was returning to them
> and feeling like I was meeting an old friend. One day I
> sat down with a particular nettle, sat in a patch of many
> others, and I felt a really strong 'star'-like quality. It is very
> hard to describe but it felt like this enormous spreading,
> shining sensation – like an expanding force of intense
> energy. I intuited it as a gesture of the wholeness of the
> plant. A wholeness that I could then recognise in parts of
> the plant such as the force of the 'sting' that you feel when
> touching the small syringe-like 'stinging hairs'; the shape
> and expression of the thousands of tiny hairs seemingly
> bursting out of the plant with this immense energy; the
> pattern of 'spikes' on the leaf edges which feel like they are

dynamically spreading outward with purpose. The whole plant felt like a star that was 'shining'. A wonderful experience to participate in.[29]

What she experiences as a gesture expressing the *quality* of the plant *is* the appear*ance* of the plant – cf. Wittgenstein on gesture. This is a beautiful example of the being of living nature 'making itself manifest to us'. Instances of this kind are invaluable because they help us to build up confidence that we can learn to encounter nature as the expression of living being. The way into this encounter begins with taking appearance seriously.

Notes

Chapter 1

1. L. von Bertalanffy, *General System Theory* (New York: Braziller, 1968), p.87.
2. Although Bohm disagreed with Bohr, he had great respect for him as a natural philosopher, and at the time in question (1962–64) he was emphasising that in order to overcome Bohr it was first necessary to really understand him. He continually pointed out the subtlety of Bohr's thinking, and how his own followers (i.e. Bohr's) often did not appreciate this sufficiently. It was a decade later before Bohm adopted the language of the implicate and explicate orders (enfolding and unfolding) as the basis for a holistic description of quantum phenomena. The two part seminal paper, 'Quantum Theory as an Indication of a New Order in Physics' was published in *Foundation of Physics* (Part A in vol. 1, no. 4, 1971; Part B in vol. 3, no. 2, 1973). These are included in David Bohm, *Wholeness and the Implicate Order* (London: Routledge and Kegan Paul, 1980). The idea of the implicate/explicate distinction, and its wider applications, is discussed very clearly in Paavo T.I. Pylkkänen, *Mind, Matter and the Implicate Order* (Berlin: Springer, 2007). See especially pp.19–20, and pp.58–59, for the whole and the part in the hologram and other examples.
3. Brian Lewis was particularly interested in the work of Hans-Georg Gadamer. The English translation of Gadamer's *Truth and Method* had not yet appeared but in the meantime we were fortunate to have available what Lewis described as 'the first readable book on hermeneutics in English': Richard E. Palmer, *Hermeneutics: Interpretation Theory in Schleiermacher, Dilthey, Heidegger, and Gadamer* (Evanston: Northwestern University Press, 1969)
4. This was at a conference on 'Developing the Whole Man', which was the twenty-fifth anniversary conference of the Institute for the Comparative

177

Study of History, Philosophy and the Sciences. It was intended to be a platform for launching a new venture in adult education, the International Academy for Continuous Education, which was the inspiration of the founder and director of ICSHPS, the mathematician and philosopher, J.G. Bennett. When he said to me, 'You should give a talk', after one of the speakers had dropped out, I was taken aback and told him that I had no idea what to talk about. He went silent, as was his custom, and after a while he said, in the *ex cathedra* manner which was also his custom, 'You should give a talk on "The Whole: Counterfeit and Authentic".' Looking back now, I am astonished at how this encapsulated the struggle that had been going on in me at the time – and which he had himself put up with at times with remarkable patience. In the event the talk was well received, and I wrote an extended version which was published in the house journal of ICSHPS, *Systematics*, vol.9, no. 2, Sept. 1971. Some years later, David Seamon suggested republishing it in a book he was co-editing on environmental phenomenology. Because this was very different from the original context in which the paper was written, I changed it by removing the second half of the paper and substituting a new section on the whole and the part in Goethe's way of science – which had not been known to me when I wrote the original paper. This seemed to fit surprisingly well, so I changed the title to 'Counterfeit and Authentic Wholes: Finding a Means for Dwelling in Nature', and it was published in David Seamon and Robert Mugeraur, eds. *Dwelling, Place and Environment: Towards a Phenomenology of Person and World* (Dordrecht: Martinus Nijhoff, 1985), and subsequently republished in David Seamon and Arthur Zajonc, eds. *Goethe's Way of Science: A Phenomenology of Nature* (Albany: State University of New York Press, 1998). Under the title 'Authentic and Counterfeit Wholes', it also forms Part 1 of Henri Bortoft *The Wholeness of Nature: Goethe's Way toward a Science of Conscious Participation in Nature* (Great Barrington: Lindisfarne Books, 1996 and 2005); published at the same time in the UK, with the subtitle *Goethe's Way of Science*, by Floris Books (Edinburgh).

5. Ingrid Lemon Stefanovic, *Safeguarding Our Common Future: Rethinking Sustainable Development* (Albany: State University of New York Press, 2000), p.89.

6. Iain McGilchrist, *The Master and His Emissary: The Divided Brain and the Making of the Western World* (New Haven: Yale University Press, 2009), p.179.

7. Referred to in note 4.

8. Perception is discussed in my previous book, *The Wholeness of Nature*, pp.50–57 and pp.123–137. Language is discussed in Chapter 5 below.

9. Richard Hamblyn, *The Invention of Clouds: How an Amateur Meteorologist Forged the Language of the Skies* (London: Picador, 2001), p.35.

10. Martin Heidegger, *Identity and Difference* (New York: Harper & Row, 1969), p.29.

11. The use of 'differences' in a verbal manner may at first cause difficulty for some readers – and the same goes for 'differencing' which will also be used. A precedent for this may be the use of 'presences' and 'presencing', which is now commonplace in discussions of Heidegger's philosophy – although there are some who continue to find this objectionable. Normal language use often focuses more on the noun than the verb, the static instead of the dynamic. But when it is the latter which needs to be emphasised, then it may be useful to introduce an unfamiliar, and therefore at first awkward, term which is more dynamic. I have noticed that, although they may look more awkward on the printed page, these more dynamic terms are usually readily accepted when spoken – in fact I often just use them as a matter of course without comment. On one occasion when I did make a comment, Nigel Topping, a mature student on the Masters in Holistic Science program at Schumacher College, pointed out that, as we can go from 'dance' to 'dancing', so we could just as easily go from 'difference' to 'differencing'. This may not satisfy the voice of the schoolmaster echoing in my ear, but I am going to ignore that because the schoolmaster doesn't always know about the needs of philosophical work.

12. Quoted in Oliver Sacks, *The Man Who Mistook His Wife for a Hat* (London: Pan Books, 1986), p.89. Charcot was pre-eminent in the field of neurology in the later part of the nineteenth century. Sigmund Freud became his pupil in Paris in the 1880s. Oliver Sacks also describes his own experience of coming to recognise Tourette's syndrome, a disease which at the time was believed to be extremely rare. He began to recognise the symptoms of this in the behaviour of people in New York, and came to the conclusion that 'Tourette's is very common but fails to be recognised, but once recognised is easily and constantly seen'. (*Ibid.* p.89.).

13. Richard E. Palmer, 'Phenomenology as a Foundation for a Post-Modern Philosophy of Literary Interpretation', *Cultural Hermeneutics,* vol. 1, 1973, p.213.

14. Günter Figal, 'Hermeneutics as Phenomenology', *Journal of the British Society for Phenomenology,* Volume Forty, Number Three, October 2009, p.257; Edmund Husserl, *The Idea of Phenomenology* (The Hague: Martinus Nijhoff, 1964), p.11.

15. Brice R. Wachterhauser, *Beyond Being: Gadamer's Post-Platonic Hermeneutic Ontology* (Evanston: Northwestern University Press, 1999), p.144.

16. McGilchrist, *The Master and his Emissary*, p.133.

17. Maurice Merleau-Ponty, *Phenomenology of Perception* (London: Routledge and Kegan Paul, 1962), p.394. The *Fundierung* relation was first introduced by Husserl in his *Logical Investigations* (1900–1901), and later taken up and developed by Merleau-Ponty – see M.C. Dillon, *Merleau-Ponty's Ontology* (Evanston: Northwestern Press, 1997), p.195.

18. McGilchrist, *op. cit.* p.230.

19. *Ibid.* p.231.

20. Martin Heidegger, *Introduction to Metaphysics* (New Haven: Yale University Press, 2000), p.107. The translation I have quoted here differs slightly, and is the one given in Michael E. Zimmerman, *Heidegger's Confrontation with Modernity: Technology, Politics, Art* (Bloomington: Indiana University Press, 1990), p.224/5.

21. The publication, towards the end of the last century, of Heidegger's early lectures and seminars, from 1919 up to the publication of *Being and Time* in 1927 – the decade during which Heidegger was at the height of his powers as a teacher – has thrown much light on Heidegger's philosophical project. Thomas Sheehan points out that Heidegger's very terminology is a barrier to understanding this project – see Thomas Sheehan, 'The Turn', in Bret W. Davis, ed., *Martin Heidegger: Key Concepts* (Durham: Acumen, 2010), p.83. The language of 'being' that he uses (being, beings, being of beings) comes from a *pre*-phenomenological tradition of the metaphysics of objective realism. This does not sit well with Heidegger's *phenomenological* understanding of being as meaning. When Heidegger refers to 'beings', he is not referring to things as just existing-out-there, but to things as appearing – and phenomenologically, to appear is to mean. There is a shift of emphasis from the metaphysical to the hermeneutic, from an entity's 'being-out-there' to its 'appearing-as', i.e. its meaningfulness. We hamper our understanding by our failure to distinguish clearly between meaning in the sense of the meaning which *is* what is seen, and meaning in the sense of the meaning *of* what is seen (cf. *The Wholeness of Nature*, pp.52–53 and pp.131–32). The meaning which *is* what is seen is the meaning which it is – i.e. being (what-is) is meaning. Woody Allen seems to have got it right: 'to adapt Woody Allen's phrase: meaning is just another way of spelling being' (Sheehan, *Ibid.*)
    In the first lecture course he ever gave, just after the First World War, Heidegger tried to make his students aware of what they directly encounter in lived experience:

> This environmental milieu (*Umwelt*) ... does not consist just of
> things, objects, which are then conceived as meaning this and this;
> rather, the meaningful is primary and immediately given to me
> without any mental detours across thing-oriented apprehension.
> (Martin Heidegger, *Towards the Definition of Philosophy*, p.61.
> Sheehan's more colourful translation will be found in Davis, ed.,
> p.83.)

Because of the primacy of meaning (which the later Wittgenstein also uncovered), appearing is appearing-as (not 'appearing as'), and concomitantly seeing is seeing-as (not 'seeing as'). Thus, as Sheehan says: 'one of the challenges in interpreting Heidegger is to remember that when he uses the language of "being", he means "being" as phenomenologically reduced, i.e. as meaningfulness'. So what 'Heidegger's phenomenological theory of being-as-meaning ... investigates is not meaningful *things* but their *meaningfulness*'. (Thomas Sheehan, 'Dasein', in Hubert L. Dreyfus and Mark A. Wrathall, eds., *A Companion to Heidegger*, p.197.) In other words, to understand Heidegger we have to cease taking 'being' metaphysically and take it hermeneutically instead.

22. Dillon, *op. cit.* p.xvii.
23. The recognition that there cannot be an absolute distinction is fundamental to Hegel's philosophy. The fact that a distinction necessarily entails a relation led Hegel to see that the principles of Aristotelian logic – especially the principles of identity and contradiction – are often interpreted in a way that is too literal and dogmatic (see Edward Caird, *Hegel*, Chapter 7). It is often forgotten that Aristotle derived the principles of deductive logic by careful observation of the thinking of mathematicians – 'Deductive logic is, in effect, the child of mathematics'(Morris Kline, *Mathematics: The Loss of Certainty*, p.21). But once abstracted from the context of its discovery, Aristotle's logic was subsequently taken to be applicable to *all* forms of reasoning. There is in fact no warrant for this universalisation of the specific form which the practice of reason took in specific circumstances. Even though the principles of identity, contradiction, and the excluded middle, may be appropriate for the kind of formal proof by deductive reasoning practised by the mathematicians of the Athenian Academy in the case of geometry and number theory, this does not mean that such principles can be assumed to apply *in the same way* elsewhere. Hegel recognised that they are too isolating for most situations, leading us into the error of treating things and thoughts as being self-identical, and hence as being separate and independent. Although they may be considered in this way for certain

purposes, it has to be remembered that there is an intrinsic relation in every distinction, and this stops everything from falling apart into atomic existences. Thus, what Hegel did was to recognise that Aristotle's principles of logic are relatively, and not absolutely, valid.

24. Charles B. Guignon, *Heidegger and the Problem of Knowledge* (Indianapolis: Hackett Publishing Company, 1983), p.100. Heidegger developed a holistic account of entities in *Being and Time* with his notion of a referential totality. According to this, any entity is what it is only in terms of its place in a web of relations that forms the context within which the entity is meaningful. As well as Guignon, pp.99–100, see Cristina Lafont, 'Hermeneutics', in Dreyfus & Wrathall, *A Companion to Heidegger*, pp.265–84.

Chapter 2

1. A very helpful account of this more comprehensive approach to the early development of modern science is given in A.C. Crombie, *Robert Grosseteste and the Origins of Experimental Science 1100–1700* (Oxford University Press, 1953), from which what follows is taken.

2. *Ibid.* p.24.

3. Robin Dunbar, *The Trouble with Science* (Cambridge, MA: Harvard University Press, 1995), p.17.

4. Crombie, p.25.

5. *Ibid.* The notion of induction in Aristotle, and later in medieval philosophy, is much richer than it is in the modern period after Hume. It was in the nineteenth century in particular that induction seems to have been restricted to the process of enumerative generalisation. Aristotle discussed two types of induction. The first, induction by enumeration, is the one with which we are familiar today. But it is the second type of induction that is crucial in science. This is intuitive induction, which is the direct intuition of meaning in the sense experience – what we would now call an 'insight'. Aristotle gives the example of someone who notices on several occasions (though in fact only one may be needed) that the bright side of the Moon is turned towards the Sun, and *sees* that the Moon shines by reflecting light from the Sun. See John Losee, *A Historical Introduction to the Philosophy of Science*, third edition (Oxford University Press, 1993), pp.7–8. A brilliant account of induction and intuition, both historically and in the context of philosophical understanding today, is given in Louis Groarke, *An Aristotelian Account of Induction: Creating Something from Nothing* (Montreal: McGill-Queens University Press, 2009).

6. Morris Kline, *Mathematics: The Loss of Certainty* (New York: Oxford University Press, 1980), p.21.

7. Crombie, p.3.

8. *Ibid.* p.318.

9. Galileo, *Dialogue Concerning the Two Chief World Systems,* trans. Stillman Drake (Berkeley: University of California Press, 1953; revised edition, 1967, p.xvii and p.xix). Galileo used the same terms for the double procedure – 'resolution' and 'composition' – that were used by Grosseteste and others four hundred years earlier, and which were used in the university of Padua when Galileo was there. These are the Latin translations of the Greek terms, 'analysis' and 'synthesis', for the double procedure – which were used by Descartes.

10. Heinz R. Pagels, *The Cosmic Code: Quantum Physics as the Language of Nature* (London: Penguin Books, 1984), p.304.

11. Hubert L. Dreyfus, *Being-in-the-World: A Commentary on Heidegger's "Being and Time", Division 1* (Cambridge MA: MIT Press, 1991), p.280.

12. *Ibid.* p.263.

13. Thomas S. Kuhn, *The Copernican Revolution: Planetary Astronomy in the Development of Western Thought* (Cambridge MA: Harvard University Press, 1957), p.142.

14. *Ibid.* p.180.

15. *Ibid.* p.131. during the ten years that Copernicus spent in Italy (1496–1506) he studied at Bologna under Domenica de Novara, who was closely associated with the Neoplatonists of the Florentine Academy (Marcilio Ficino and Pico della Mirandola) which was sponsored by the Medici family. See E.A. Burtt, *The Metaphysical Foundations of Modern Physical Science,* second ed. (London: Routledge and Kegan Paul, 1932), p.54. This brilliant book is extremely useful for the kind of background to the development of modern physical science with which we are concerned here.

16. Quoted in Brian Easlea, *Witch Hunting, Magic and the New Philosophy: An Introduction to the Debates of the Scientific Revolution 1450–1750* (New Jersey: Humanities Press, 1980), p.59.

17. See the quotation from Kepler in Burtt, *op. cit.* p.59.

18. Arthur Koestler, *The Sleepwalkers: A History of Man's Changing Vision of the Universe* (London: Penguin Books, 1959), p.264.

19. This two-world dualism is often just called 'metaphysics' in modern European philosophy. The question is, to what extent did Plato himself introduce a separation between the intelligible and the sensible, which then crystallised into the theory of two worlds in what has been called the

'standard traditional interpretation' of Plato? Gadamer calls this 'the vulgar conception of Platonism' and refers to it as pseudo-Platonism. He sees one source of the misunderstanding here coming from the failure to take into account sufficiently the remarkable step which Plato took in distinguishing clearly between mathematical and empirical thinking. Before Plato it seems that mathematics was seen as being an empirical kind of activity – in the case of geometry, for example, it was seen as depending on measurements made on figures that were drawn, such as circles and triangles. Plato's achievement was to show that what is truly *mathematical* does not depend on working from sensory images of geometrical figures – for example, the discovery that the sum of the interior angles of a triangle is equal to two right angles (180°) does not depend on measuring the angles of drawn triangles, but follows directly from the very idea of a triangle. It was to convey this – which many at the time found difficult to comprehend (and still do) – that Plato introduced the *distinction* between the sensible and the intelligible aspect of things:

> Such a distinction is particularly essential to mathematics if mathematicians are to understand their own thought by its own inherent standards and not mistake it for a type of empirical research, a temptation to which even a gifted mathematicians like Thaetetus was susceptible. (Wachterhauser, *Beyond Being: Gadamer's Post-Platonic Ontology*, p.82.)

This means keeping these two aspects *separate* in the *practice* of mathematical thinking:

> Gadamer contends that one key motivation that Plato had for positing the *chorismos* (separation) was to delineate in an adequate fashion for the first time the basic elements of an ontology of mathematics. This was necessary for the advancement of the flourishing mathematical enlightenment in Greece at the time, which Plato experienced, supported and to which he perhaps even contributed. (*Ibid.* See Gadamer, *The Idea of the Good in Platonic-Aristotelian Philosophy*, pp.16–18.)

This separation is entirely methodological – as we see so clearly when we put it back into the situation in which it was first made. But this *methodological distinction* was not intended to be taken as an *ontological separation* into two worlds. Yet this confusion is what seems to have happened, and the resulting misunderstanding, which we call 'Platonism', considers that there is 'a second world, supposedly separated from our world of appearances by an ontological hiatus' (Gadamer, *The Idea of the Good,* p.16). But if this were the case, the ideas would exist as such entirely independently of appearances. However, Gadamer points out that 'the ideas are ideas *of* appearances and that they do not constitute a world existing for itself', and consequently 'the complete separation of a world of the ideas from the world of appearances would be a crass absurdity' *(Ibid.).* We should think rather of a unitary structure which is sensible/intelligible, in which the intelligible is also an aspect of this world, and not another world separated from a world of appearances. Gadamer's fundamental point that the ideas are the ideas of appearances, means that the appearances themselves have an intelligible as well as a sensory side:

> Thus there is a fundamental connection between appearance and Ideas that enables Gadamer to speak of them as two sides of the same reality. If this is so, then there is no two-world ontology that must be attributed to Plato. (Wachterhauser, *op. cit.* p.83.)

According to the 'standard traditional interpretation', the two-world theory of Ideas (or Forms, as they are also sometimes called) is at the centre of Plato's philosophy. This is first mentioned in *Phaedo,* and given its most explicit statement in the *Republic.* After this, so the story goes, Plato began to become increasingly aware of difficulties with this two-world theory, and in the later dialogue, *Parmenides,* he subjected his own theory to severe criticism, which led him to modify it considerably or even to reject it. This is not the way that Gadamer understands it. Based on his lifelong experience of reading and teaching Plato, he came to think that it should be the other way round. Far from representing Plato becoming critical of his own philosophical innovation, what the *Parmenides* actually shows us is Plato trying to correct a *misunderstanding* of the theory of Ideas that had developed, namely the two-world theory itself. Although it is unlikely that Plato himself ever held such an absurd theory, it is nevertheless possible that he unintentionally contributed to it by the way that he first introduced the Ideas. Realising this, he set out to correct this false interpretation, which is what he does in the *Parmenides,* where he pushes us into a complete separation of the world of Ideas from the world of appearances 'precisely in order to reduce such an understanding ... to absurdity (see *Parmenides* 133b ff)' (Gadamer, *Ibid.)*

> Thus, for Gadamer, these later dialogues represent the culmination of Plato's metaphysical inquiries and not Plato's self-critical dismantling of his earlier work. Instead these dialogues provide decisive clues for interpreting his earlier works. (Wachterhauser, *op. cit.* p.5.)

It is ironic that Plato has always been identified with the two-world theory of Ideas and appearances, a bifurcation which he himself rejected, and even more ironic that his own attempt to correct this misinterpretation has so often been seen as his criticism and rejection of a theory which he probably never held in the first place.

20. G.B. Madison, *The Hermeneutics of Postmodernity: Figures and Themes* (Bloomington: Indiana University Press, 1988), p.105 and p.106.

21. Quoted in Burtt, *op. cit.* p.79.

22. Quoted in Ernst Cassirer, *Substance and Function and Einstein's Theory of Relativity* (New York: Dover Publications, 1953), p.371. Cassirer does not mention Galileo in this context, or Goethe's admiration for the way that he exemplified this principle, but he introduces it in connection with Einstein's postulate of the constancy of the measured speed of light which becomes the basis of the Special Theory of Relativity.

23. The radically new conception of motion that Galileo introduced – which entailed separating the motion that a body has from the nature of the body itself, so that a body is *indifferent* to its state of motion – is described in Richard S. Westfall, *The Construction of Modern Science: Mechanisms and Mechanics* (Cambridge: Cambridge University Press, 1977), pp.19–21.

24. Quoted in Westfall, p.17.

25. *Ibid.* p.21.

26. The philosophy of atomism was originally developed in Greek philosophy (an atomic philosophy was also developed in India) by Democritus and Leucippus in the pre-Socratic period as a response to the arguments of Zeno (following Parmenides) that change is impossible. The philosophy of atomism was taken up again in the period after Plato and Aristotle by Epicurus. But this time it was used as the basis for a way of life. The physics of Epicurus was based on the idea that what is *real* is an infinite space (void) with an infinite number of atoms in random and unceasing motion. What was important for Epicurus and his followers about this physics is that it led to an ethics which, by following certain practices, would lead the individual into a state of freedom from fear and anxiety. This was its 'selling point' when it spread eventually from the Greek to the Roman world. The Epicurian physics and the ethics it supported were given a full account by

the Roman poet Lucretius in *The Nature of the Universe (De Rerum Naturae)* in the first century AD. The manuscript of this poem was rediscovered in 1417. At first the interest was primarily in the ethics as a practical way of life. But in the Epicurian philosophy this depends on the physics, and eventually the physics was abstracted and used on its own as a mode of explanation for physical phenomena – it fitted the emphasis on number very well. It could take different forms – strictly atomic or just generally corpuscularian, for example – but these differences do not concern us here. See Thomas S. Kuhn, *The Copernican Revolution*, p.237 *seq.* for a brief discussion of the corpuscularian philosophy in physics and its influence on the direction taken by the investigation into the physics of motion in the seventeenth century.

27. The complete text is in Stillman Drake, *Discoveries and Opinions of Galileo* (New York: Doubleday, 1957). This passage is on p.274.

28. Burtt, *op. cit.* p.90.

29. Descartes in a letter of October 1638, quoted in Tom Sorell, *Descartes* (Oxford: Oxford University Press, 1987), p.2.

30. Westfall, *op. cit.* p.38. Although Descartes was primarily a mathematician and mathematical visionary, he is more usually thought of as the 'father of modern philosophy' because the mathematical context of his philosophy is ignored and he is presented as 'purely' a philosopher. He was convinced that the mathematical style of thinking could replace Aristotelian philosophy in the synthesis with Christianity that Aquinas had produced in the Middle Ages, and which had become the official doctrine of the Church. This was his ultimate ambition, but he thought the place to start was with the more limited problem of providing a philosophical foundation for the then new science of mathematical physics. It was this foundation which he set out to provide, firstly in his *Discourse on the Method of Properly Conducting one's Reason and of Seeking the Truth in the Sciences (1637),* written for the generally educated reader, and subsequently in *Meditations on the First Philosophy in which the Existence of God and the Real Distinction between the Soul and the Body of Man are Demonstrated* (1641), written for the theologians in Paris.

31. Sorell, *op. cit.* , p.3.

32. Westfall, p.39.

33. David E. Cooper, *Existentialism: A Reconstruction* (Oxford: Blackwell, 1990), p.23.

34. Sorell, p.84.

35. Kurt Hübner, *Critique of Scientific Reason* (Chicago: University of Chicago Press, 1983), p.125.

36. *Ibid.* p.130.

37. *Ibid.* p.131.

38. See Margaret Wertheim, *Pythagoras' Trousers: God, Physics, and the Gender Wars* (London: Fourth Estate, 1997), Chapter 4, for a very useful summary.

39. Westfall, *op. cit.* p.31.

40. In view of this historical background, it is surprising how often the 'subject-object dichotomy' is attributed to Descartes. But this really makes no sense, because this dichotomy is an inevitable consequence of the dynamics of cognitive experience itself. If Descartes had been a fishmonger instead of a mathematician and philosopher, there would still be the subject-object dichotomy. More importantly, if the historical circumstances had been otherwise than they were, there would have been no occasion for such an extreme form of dualism to be put forward by Descartes or anyone else. But there would still be the subject-object dichotomy because subject and object precipitate out together from the cognitive experience. Whenever we focus attention explicitly on seeing-knowing the world, the outcome will be a sense of 'myself' as separate from what is seen, which stands over against me as an object (the German word *gegenstand* conveys this precisely). If we look specifically at a cup, a clock, or whatever our gaze happens upon, we will find that the outcome is a sense of ourselves standing back – and hence separate – from what we are now looking *at*. There is the sense of an observer separate from an object. There is no such separation if we can catch seeing 'in the act' – i.e. in the lived experience. It only appears when attention is focused on *what* is seen instead of the *seeing* of what is seen. But this is where we begin when we give a 'common sense' description of experience, and consequently we reflexively project the subject-object dichotomy back into the lived experience where it doesn't belong. With the emphasis on cognition that has grown in the modern world, especially as a result of interest becoming focused on scientific knowledge, it is understandable that the subject-object dichotomy has come to occupy a central place in discussions about knowledge and our cognitive relation to the world. Surprisingly, it turns out that the cognitive mode is not in fact the primary way in which we are engaged with the world. This is one of the remarkable discoveries to have come out of phenomenology – specifically the existential approach to lived experience taken by Heidegger in the 1920s. As a consequence, the subject-object dichotomy no longer has the prominence that it had when the focus was on cognition as the primary mode of engagement with the world.

41. Oliver Sacks, *The Man Who Mistook His Wife For A Hat* (London: Pan Books, 1986), Chapter 3. All the quotations following are taken from this chapter.

42. Dan Zahavi, *Husserl's Phenomenology* (Stanford: Stanford University Press, 2003), p.21. Sokolowski indicates the reason why the radical nature of Husserl's breakthrough often fails to be appreciated:

> Husserl made a decisive breakthrough in modern thought:
> he showed the possibility of avoiding the Cartesian, Lockean
> concept of consciousness as an enclosed sphere; he restored the
> understanding of mind as public and as present to things. He opens
> the way to a philosophical realism and ontology that can replace
> the primacy of epistemology. Many of these positive possibilities in
> Husserl's thought have not been appreciated because the Cartesian
> grip – "la main morte de Descartes" – is so strong on so many
> philosophers and scholars. All too frequently, everything in Husserl
> is reinterpreted according to the very positions he rejected. (Robert
> Sokolowski, *Introduction to Phenomenology*, p.226.)

43. Although the second volume of *Ideas* was written at the same time as the first volume (which was published in 1913), it was not published in Germany until 1952, fourteen years after Husserl's death, and an English translation was not published until 1989. It is now clear that the extraordinary range and depth of Husserl's phenomenology was underestimated for so long simply because so little of his work had been published, and also that the standard interpretation of Husserl is often misleading for the same reason. Merleau-Ponty was able to study the manuscript of the second volume of *Ideas* in the newly opened Husserl Archive in Belgium in 1938, and this forms the basis for his own philosophy of the lived body – see David R. Cerbone, *Understanding Phenomenology* (Chesham: Acumen, 2006), pp.98–105.

44. See, for example, Ted Toadvine, *Merleau-Ponty's Philosophy of Nature* (Evanston: Northwestern University Press, 2009)

45. Alexandre Koyré, *From the Closed World to the Infinite Universe* (Baltimore: Johns Hopkins University Press, 1957), p.176.

46. *Ibid.* p.178. The quotations are from the second and third letters respectively.

47. I. Bernard Cohen, *The Newtonian Revolution: with illustrations of the transformation of scientific ideas* (Cambridge: Cambridge University Press, 1980.

48. Dunbar, *op. cit.* p.42.

49. Roger Bacon, quoted in Crombie *op. cit.* p.143.

50. The dictionary offers three adjectives: sensory, sensual, sensuous. Something more than 'sensory' is needed in order to distinguish the emphasis on the senses in Goethe's approach to nature from the senses in the more familiar

empirical approach. 'Sensuous' can have unwanted connotations because it is often conflated with 'sensual'. But, if this can be avoided, it is a better word to use than 'sensory' in order to draw attention to the way that Goethe's approach is nevertheless different from mainstream empiricism. A precedent for this use is in the paper by Thomas R. Blackburn, 'Sensuous – Intellectual Complementarity in Science,' in Robert E. Ornstein, ed., *The Nature of Human Consciousness: A Book of Readings* (San Francisco: W.H. Freeman and Co., 1973), pp.27–40.

51. In this experiment, first done by Jerome Bruner and his colleagues in the 1950s, the participants are shown a sequence of images of playing cards on a screen. Unknown to them, several anomalous cards are included – for instance, a black 'Five of Hearts', or a red 'Ace of Spades', and so on. What emerged from this experiment was very surprising at the time. In the two examples just given, they would *see* the 'Five of Spades' or the 'Ace of Hearts'. The anomaly was not noticed at first. It was not a case of first seeing a card which seemed odd and then interpreting it as a normal card. A normal card was seen directly. What actually occurred was replaced by the category, so that in this case we could say that the participants *saw* the category and not the actual occurrence. When the time for which each card was on the screen was gradually increased, there came a point where the anomaly began to obtrude on awareness. This realisation turned out to be unexpectedly stressful for the participants. As their experience became ambiguous – before there was a clear recognition of the 'trick' that had been played on them – some responded with panic, and some with anger. Although this response may seem excessive to the outsider, it becomes understandable when we realise that, to the participants, it seemed as if their very experience of seeing was somehow being tampered with, so that they could not tell at first whether what was happening was 'out there' or 'inside their own heads'. Only when the anomalous cards were left on the screen for a longer time did it become clear to them what had been going on.

52. Heidegger's distinguinction between *belonging* together and belonging *together* has been introduced in Chapter 1, note 10. The sensuous-intuitive mode of perception leads us to the *belonging* together which is the wholeness of the phenomenon, whereas the verbal-intellectual mind is more at home with the belonging *together* of the system.

53. The motivation is completely different in the two cases. Goethe's motive was to understand the qualities of colour, and hence his science of colour is the science of these qualities as such. Newton's motive, on the other hand was to eliminate *unwanted* colour in optical instruments. This is

really a branch of mathematical-instrumental optics and does not require us to enter into the *experience* of colour. Contrary to what is often said, there is no disagreement between Goethe and Newton once the context of the motivation is taken into account. Goethe himself eventually came to understand this, but unfortunately others were less comprehensive in their understanding. Goethe's approach is described very clearly in detail in Heinrich O. Proskauer, *The Rediscovery of Colour: Goethe versus Newton Today* (Spring Valley: Anthroposophic Press, 1986), which includes a prism and sixteen black and white and multicoloured plates for the reader to make the observations discussed in the book for herself. See also Henri Bortoft, *The Wholeness of Nature*, part 3, Chapter 4.

54. Günter Theissen and Heinz Saedler, 'Plant Biology: Floral quartets', *Nature*, 409, 469–71 (25 January 2001)

55. Martin Heidegger, *Being and Time*, translated by John Macquarrie and Edward Robinson (London: SCM Press, 1962), p.191.

56. Johann Wolfgang von Goethe, *The Metamorphosis of Plants*, introduction and photography by Gordon L. Miller (Cambridge MA: MIT Press, 2009).

57. Iain McGilchrist, *The Master and His Emissary: The Divided Brain and the Making of the Western World*, p.1.

58. *Ibid.* p.4.

59. *Ibid.* p.179.

Chapter 3

1. Chapter 2, note 54. See Brian Goodwin, *How the Leopard Changed its Spots: the Evolution of Complexity* (London: Weidenfeld and Nicolson, 1994), pp.120–25.

2. The expression 'learning to think like a plant lives' is due to Craig Holdrege. Goethe's practice of active seeing and exact sensorial imagination is described in the previous chapter.

3. The English translation of *Die Metamorphose der Pflanze* which I am using is the one which appeared in the *Journal of Botany* in 1863. It is available from the Biodynamic Farming and Gardening Association (Kimberton, 1993). A much more recent translation is given in Douglas Miller, ed., *Goethe: Scientific Studies*, (New York: Suhrkamp, 1988), pp.76–97. My choice of translation is simply determined by the fact that it is the one with which I am most familiar because I have used it in lectures and workshops. There are no major divergences between the two translations in the quotations I use. The major difficulty for anyone trying to read *The Metamorphosis of Plants*

is the fifty or so different plants which Goethe mentions in the text. For the reader who lacks familiarity with these, or does not have a suitable botanical guide to hand, this makes reading Goethe's text rather a dry experience. What may have been familiar in Goethe's time is forgotten about today. As a result, up until recently, what has been available has been no more than an abstract description. But now this has changed. Gordon Miller has prepared an edition, based on Douglas Miller's translation, in which he has included photographs of the plants that he has taken, situated in the text at the places where Goethe refers to them. See *The Metamorphosis of Plants,* introduction and photography by Gordon L. Miller (Cambridge MA: MIT Press, 2009). The result is the transformation into a living work of what otherwise is no more than a skeleton. The beauty of the photographs taken with such care is breathtaking. We can only be astonished at Goethe's detailed observational work, together with the overall vision of the idea of metamorphosis, which biology today recognises as the truth of the plant. Certainly anyone who harbours the thought that Goethe was merely a dilettante playing at science should look for themselves at the painstaking detail of this work (the same could be said of his work with colour). It seems that Goethe always hoped that eventually it would become possible to provide pictures of the plants where they are mentioned in the text. Well now it has been done, and we have the complete book for the first time. I cannot help feeling that it would delight Goethe, as it will all those readers for whom it makes Goethe's work more accessible than ever before.

4. This diagrammatic representation of intermediate stages between petal and stamens in the white water lily is taken from Gerbert Grohmann, *The Plant, vol 1: A Guide to Understanding its Nature* (London: Rudolf Steiner Press, 1974), p.43. There is also a diagram in *The Metamorphosis of Plants* (2009), p.44.

5. Although we will not go into this – because we are concerned here with the idea of metamorphosis, and not with the details of plant growth as such – the central organ of the pistil, which includes the ovary, is included in this transformation. For example, it can happen that there is a retrogressive step where stem leaves appear in the place of sepals (the leaves which form the outer casing enclosing the floral bud) – there is a photograph of a dandelion in which this can be seen in Grohmann, *op. cit.* p.63. But more striking than this is the case of retrogression which Goethe called a proliferous carnation (para.105 of *Metamorphosis*). In this case he observed a carnation in which the seed capsules of the ovary were transformed back into sepals (notice how easy it is to convey the false idea that this is a physical transformation), and in place of the seed capsules a second flower grew out of the first.

6.   Goethe's work on metamorphosis extends beyond the individual plant, to include variations in a plant species, members of a family of plants, such as the *Rosaceae* for example, and ultimately to the plant kingdom as a whole. The last two quotations from Goethe refer to different plants, whereas so far we have considered only the organs of a single plant. However it will not be difficult to recognise that this is a natural extension to make, and we will be considering it below. In any case, it is easy to see how both these quotations apply to the different organs of a single plant equally as well as to different plants.

7.   Rudolf Steiner, *Goethe's World View* (Spring Valley: Mercury Press, 1985), p.81.

8.   F.W.J. Schelling, *First Outline of a System of the Philosophy of Nature* (Albany: State University of New York Press, 2004), p.15. First published in 1799, when Schelling was twenty-four, it built upon his earlier work, *Ideas for a Philosophy of Nature*, published in 1797 (and revised in 1803). An English translation of the latter was published by Cambridge University Press in 1988. These are the first English translations of these works to be published. The interaction between Goethe and Schelling and their mutual influence is described in detail in Robert J. Richards, *The Romantic Conception of Life: Science and Philosophy in the Age of Goethe* (Chicago: University of Chicago Press, 2002). However, it needs to be emphasised that, notwithstanding their fruitful interaction, Goethe's way of science is very different in practice from the approach to nature developed by Schelling and others (notably Hegel) at the time, which was called *Naturphilosophie* or 'nature philosophy'. This was a development of the post-Kantian philosophy which is usually, if misleadingly, called 'German Idealism', which in this case takes a transcendental approach (in Kant's sense of the term) to find the 'conditions for the possibility' of nature. This is very different from Goethe's phenomenology of nature.

9.   Gilles Deleuze, *Bergsonism* (New York: Zone Books, 1991), p.42.

10. In the first chapter we have seen that an act of distinction is a unitary act of {differencing/relating}. So in the first place a distinction differences/relates and does not separate – as it seems to do when we begin downstream with what has been distinguished already. An intensive distinction takes the same form as one that is extensive, only now it is a unitary act of {self-differencing/self-relating}, so that an intensive distinction self-differences/self-relates.

11. This will not work with the kind of holograms that can now be obtained commercially because they are manufactured by a different process, and anyone who tries it today will be both disappointed and annoyed.

The division process could be done as described with transmission holograms when these were first developed.

12. John Seymour, *The Countryside Explained* (London: Faber and Faber, 1977), p.116.

13. *Ibid.*

14. This expression was used by David Bohm.

15. This electron micrograph was very kindly given to me by Dr Bruce Kirchoff when I visited the University of North Carolina (Greensboro) in November 1996, to give a talk on 'Goethe's Science of the Wholeness of Nature'.

16. Ronald H. Brady, 'Form and Cause in Goethe's Morphology', in Frederick Amrine, Francis J. Zucker, and Harvey Wheeler (eds.), *Goethe and the Sciences: A Reappraisal* (Chicago: University of Chicago Press, 1987), p.286.

17. *Ibid.* p.287.

18. An excellent discussion of varieties, with carefully observed examples, is given in Craig Holdrege, *Genetics and the Manipulation of Life: The Forgotten Factor of Context* (Hudson: Lindisfarne Press, 1996), Chap.1.

19. Deleuze, quoted in Todd May, *Gilles Deleuze: An Introduction* (Cambridge: Cambridge University Press, 2005), p.60. See Gilles Deleuze, *Nietzsche and Philosophy* (New York: Columbia University Press, 1983; first published 1962), p.24.

20. Rudolf Steiner, *A Theory of Knowledge Based on Goethe's World Conception,* (New York: Anthroposophic Press, 1968), p.88.

21. Holdrege op.cit. (1996), p.46.

22. Darwin was deeply impressed, overwhelmed even, by the ubiquity of variation. Before he did his work with barnacles, Darwin had believed that variation is the exception in nature, occurring only in times of crisis. His barnacle work changed that. Here he found that there are no unvarying forms, and that barnacle species are, as he put it, '*eminently* variable'. What made the work of classification so difficult was that 'Every part "of every species" was prone to change; the closer he looked, the more stability seemed an illusion'. (Adrian Desmond and James Moore, *Darwin*, p.373.) Barnacles, he told Hooker, are infinitely variable; and in the context of his theory of what he called 'the transmutation of species', he went further to see variations as incipient species. There is a switch in gestalt here, like the reversing cube: in one perspective the phenomenon appears as the variations of a species, whereas in another perspective the very same phenomenon appears as the initial stages of new species. Goethe and Darwin both encountered the organism's 'potency to be otherwise' which is the self-differencing dynamic of life. But whereas Goethe saw this unceasing

variation phenomenologically, so that he *understood* it as the expression of life itself, Darwin wanted to *explain* it (in this regard he thought more like a physicist). He eventually 'found' an explanation in the key to the success of Victorian capitalism: the division of labour. Prompted by the idea of the 'physiological division of labour' put forward by the French zoologist Henri Milne-Edwards (Desmond and Moore, p.394 and p.241), and the considerable experience of his wife's family (the Wedgewoods) in the assembly-line manufacture of pottery, Darwin applied the metaphor of the division of labour to see Nature as a 'workshop' – Nature's 'manufactory of species' – in which variation produced greater functional diversity of species, so that overcrowding did not necessarily result in direct competition for food and other resources. Thus species with small functional differences could all be supported in the same area without open competition by occupying different niches for which they were each functionally adapted in their own specific way, with the result that: 'Just as a crowded metropolis like London could accommodate all manner of skilled trades each working next to one another, yet without any direct competition, so species escaped the pressure by finding unoccupied niches in Nature's market place'. (Desmond and Moore, p.420). The Malthusian problem of overpopulation and competition was solved in Nature, it seemed, in much the same way that it had been in nineteenth century industrial Britain.

23. The reason why the two different terms, Form and Idea, are often used in connection with Plato's philosophy, follows from the fact that in Greek two different words were used: *eidos* and *idea*. When these were translated into Latin they became *forma* and *idea,* from where they entered into English as 'form' and 'idea'. It is customary to write 'Idea', rather than 'idea', when dealing with Plato, to emphasise that in this context it does not mean an abstract mental idea in the modern subjective sense.

24. We saw in note 19 of Chapter 2 that the key dialogue here is the *Parmenides*. The conventional view is that this dialogue represents Plato facing up to the difficulties with the theory of Ideas and becoming self-critical, which leads him, if not to reject it, at least to a revision of the theory which presents it in a much diluted form. But Gadamer strongly disagrees with this conventional view of the *Parmenides*. He sees it as a mature work representing the culmination of Plato's thinking, and not as a rejection of his own earlier work. It is Plato's attempt to correct the mistaken interpretation of the Ideas, which he may himself have unwittingly encouraged but never intended. Gadamer also believes that the theory of Ideas is not the real core of Plato's philosophy, as usually believed, but that the true focus is 'the one

and the many', which is a necessary condition for understanding the Ideas in the right way. A very clear discussion of Gadamer's understanding in this regard is given in Wachterhauser, *Beyond Being: Gadamer's Post-Platonic Hermeneutic Ontology.*

25. It may even be that Goethe's dynamical thinking of 'the one and the many' could lead us nearer to what Plato was really trying to say.

26. Dennis Klocek, in a talk given at Rudolf Steiner College in Waterville, Maine, July 1998.

27. Goethe described this experience in his review of Purkinje's *Sight from a Subjective Standpoint* (1824). See Douglas Miller, ed., *Goethe: Scientific Studies*, p.xix. The dynamical quality of Goethe's perception of organic nature is strongly emphasised by Rudolf Steiner in *A Theory of Knowledge Based on Goethe's World Conception*, Chapter 16, especially pp.90–93. Goethe once suggested that the way in which multiple patterns can be produced with the kaleidoscope could be used as a metaphor for his dynamic experience of seeing the plant – and he was well aware of the danger of taking such a metaphor too seriously and turning it into a 'model'. Keeping this caution in mind, we can now offer the technologically updated metaphor of the multiple hologram for this experience of seeing the multiply unfolding plant.

28. We can also see this in the light of Hegel's conception of the *concrete* universal – as distinct from the more familiar abstract universal – which Hegel believed was the great advance he had made upon previous philosophers. Whereas the abstract universal is reached by *excluding* all differences, and so contains only what is common, the concrete universal is 'the universal which differentiates or particularises itself and yet is one with itself in its particularity' (Caird, *Hegel*, p.135). We must be careful not to loose sight of the fact that the concrete universal is *intrinsically* dynamic – it 'particularises itself' – otherwise we will simply reduce it to an inclusive form.

29. Peter J. Bowler, *Evolution: The History of an Idea* (Berkeley: University of California Press, 1989), p.132.

30. Adrian Desmond, *The Politics of Evolution: Morphology, Medicine and Reform in Radical London* (Chicago: University of Chicago Press, 1992), p.368.

31. Owen said that the general vertebrate 'unity of plan', as it was often called, pointed towards a 'predetermined pattern, answering to the "idea" of the Archetypal World in the Platonic Cosmogony' (quoted in Adrian Desmond, *op. cit.* p.364). The Platonic influence in comparative anatomy in Britain at this time stemmed initially from the poet and philosopher Samuel Taylor Coleridge. He had been very much influenced by German *Naturphilosophie*, which he saw as a way to combat the prevailing mechanico-corpuscular

philosophy in Britain that to him was the cornerstone of the materialism he abhorred so much. Coleridge attracted the surgeon Joseph Henry Green as his leading medical disciple, who had himself been educated partly in Germany, and had studied philosophy in Berlin. Green's protégé in turn was Richard Owen. These, among others, formed the influential group which John Stuart Mill called the 'Germano-Coleridgeans', whose aim was to promote transcendental morphology as a science of Platonic 'archetypes' existing in the Divine Mind. The morphological archetypes, therefore, became the Thoughts of God. So in this respect the science of morphology became similar to the science of mathematical physics, as this was developed in the seventeenth century by Galileo, Kepler, Descartes, and Newton. In the latter case the influence of the two-world theory led to the idea that there are mathematical laws of nature which are separate from, and on a higher ontological level than, the empirical phenomena encountered by the senses. In that case it is the mathematical laws – which function as the equivalent of the 'archetypes' in biology – which are considered to be Thoughts in the Mind of God. The development of these ideas in nineteenth century biology is described in rich detail in the work by Adrian Desmond referred to above (see especially Chapter 8)

32. Robert J. Richards, *The Romantic Conception of Life: Science and Philosophy in the Age of Goethe*, p.443.

33. *Ibid*, p.454.

34. The striking exception to this is the philosopher Rudolf Steiner. His writings on Goethe's science are saturated with the dynamic approach, so much so that, although by no means always easy to read, anyone who takes the trouble to become familiar with them can scarcely avoid beginning to pick up the experience by osmosis. This aspect of Steiner's work is much less widely known than his later work with what he called 'Anthroposophy', a more esoteric enterprise which, important as it is to his followers, has had the effect of taking attention away from other aspects of his work, especially his luminous contribution to understanding Goethe – to which anyone who wants to reach a deeper understanding will surely become indebted.

While Steiner was still a student at the Technical High School in Vienna, in 1883 when he was only twenty-one years old, he was invited to edit Goethe's scientific writings for inclusion in a collected work of German masterpieces. This took him some time to complete, the last of four volumes not appearing until 1897. This is partly because during this time he also became involved with collecting Goethe's scientific writings for the definitive Weimar edition of Goethe's works, and partly because he was also engaged in writing philosophical works of his own arising out of the problems left by Kant's philosophy – which at the time had once again become very influential. After he had finished the first of the four volumes, Steiner felt that he needed to clarify the epistemological basis of Goethe's approach to science, since this differed so much from the kind of science to which most people were accustomed. Accordingly, he wrote a short book, *Grundlinien einer Erkenntnistheorie der Goetheschen Weltanshauung*, which was published in 1886. The English translation, *A Theory of Knowledge Based on Goethe's World Conception*, is referred to in note 20 of this chapter. Chapter 16 of this work strongly emphasises the dynamic quality of Goethe's thinking, albeit in language which specifically seeks to overcome the epistemological limitation that Kant had placed on our ability to develop a science of life itself. In the same year that the fourth volume of Goethe's scientific writings was published, 1897, Steiner published another book, *Goethes Weltenshauung*, which first appeared in English translation as *Goethe's World View* (referred to in note 7 of this chapter), and has now been republished in a new translation as *Goethe's Conception of the World* (Kessinger Publishing Company, 2003). A particularly interesting feature of this book is the first part which, beginning with Plato, considers Goethe's place in the development of western thought. In the four volumes of Goethe's scientific writings that Steiner prepared for publication, he also included several introductions which he had written on various aspects of Goethe's contribution to science. These introductions were later collected together in a single volume and an English translation, *Goethe the Scientist*, was published in 1950. A new translation has now been published as *Nature's Open Secret: Introduction to Goethe's Scientific Writings* (Great Barrington: Anthroposophic Press, 2000). The dynamical approach is also evident in several of the papers collected together in David Seamon and Arthur Zajonc, eds. *Goethe's Way of Science*, referred to in Chapter 1, note 4.

35. McGilchrist, *The Master and his Emissary*, p.3 and p.33. This distinction between 'what' and 'how' is central to McGilchrist's approach to the bimodal brain, and is what enables him to 'rescue' this from the restriction of the previous either/or approach to the function of the different hemispheres of the brain.

36. *Ibid.* p.49.

37. *Ibid.* p.53 and p.54.

Chapter 4

1. E.D. Hirsch, Jr., *Validity in Interpretation* (New Haven: Yale University Press, 1967), p.46. Hirsch is an American literary critic and theorist, who was the first to write a full-length work on hermeneutics in relation to literary interpretation. However, he takes a restricted view of what hermeneutics is, limiting it, as his title suggests, to the methodology of finding valid interpretations of texts. In this respect he is more in line with the early attempts to see hermeneutics as the methodology appropriate for the human studies and social sciences. This is in contrast to the more comprehensive approach of philosophical hermeneutics which is concerned primarily with understanding as a universal dimension of human experience – as developed especially by Hans-Georg Gadamer in his magnum opus, *Truth and Method,* second, revised edition (London: Sheed and Ward, 1989). Hirsch is particularly opposed to Gadamer's hermeneutics, which he believes leads inevitably into the cul-de-sac of relativism. This is very much a consequence of Hirsch's own presuppositions, which lead him to seriously misunderstand Gadamer, whose philosophy is much more subtle than his detractors suppose – and certainly does not lead to relativism when properly understood. This will become clearer below. Gary Madison has written a very illuminating essay, 'A Critique of Hirsch's Validity', in G.B. Madison, *The Hermeneutics of Postmodernity: Figures and Themes* (Bloomington: Indiana University Press, 1988), pp.3–24, which brings out clearly the limitation of Hirsch's approach.

2. Hirsch, *op. cit.* p.44.

3. *Ibid.* p.137.

4. Madison, *op. cit.* p.19.

5. Even though this is the inevitable consequence of this direction of thinking, no one is prepared to leave it like that. The problem of how we could understand the meaning of a work is in some ways similar *in form* to the equivalent problem in the epistemology of science, namely the problem of how we can understand nature. The similarity arises from the fact that in both cases the subject-object framework is presupposed. So the problem is really that of how we can understand the meaning of a text, or how we can understand the natural world, *within the framework of the subject-object separation?* Hirsch in fact takes an approach to the problem of interpretation which is 'scientific' in the style of the physical sciences – which in effect denies that there is any difference at all between the human studies and the natural sciences, the *Geisteswissenschaften* and the *Naturwissenschaften.*

Just as in physics, where there may be several alternative hypotheses which explain the facts, and the physicist will try to devise experimental procedures to find which one is more probably true than the others, so Hirsch sees the literary critic playing an equivalent role by trying to determine which of several interpretations of a work is the one which is most probably valid – i.e. corresponds to the author's intention. But notice that all that can be identified here is the most probable meaning of the work – 'An interpretive hypothesis is ultimately a probability judgement that is supported by evidence' (Hirsch, p.180; Madison, p.5). As in the case of science, we cannot reach certainty but only what we judge to be most probable, and the reason why this seems to be so in both cases is the assumption that the object is entirely transcendent to the subject as a consequence of their separation. Having thus made the meaning of the text effectively inaccessible, Hirsch concludes that 'the root problem of interpretation is always the same – to guess what the author meant' (Hirsch, p.207; Madison, p.5). This is precisely the conclusion to which Popper came in the case of scientific knowledge, where nature is similarly made ultimately inaccessible as a consequence of separating and detaching subject from object. But such an approach to understanding the meaning of a work, which ultimately denies us access to that meaning, is surely unsatisfactory – especially when we can see that it is simply a consequence of the assumption that understanding the meaning of a work must take place within the framework of the subject-object dichotomy.

6. Robert Sokolowski, *Introduction to Phenomenology*, (Cambridge: Cambridge University Press, 2000), p.226.

7. See Chapter 1, notes 13 and 14.

8. Cf. Heidegger: 'Being means appearing. Appearing is not something subsequent that sometimes happens to being. Being presences *as* appearing'. See Chapter 1, note 20. What is so useful about Heidegger is the way that he offers a description of experience which does not presuppose that experience must be constrained by the subject-object framework. Statements which at first appear abstract, or even just 'metaphysical', turn out to be concrete descriptions of experience once this framework is suspended by shifting the position of attention in experience away from what has happened into the happening itself.

9. Jonathan Lear, *Aristotle: the desire to understand* (Cambridge: Cambridge University Press, 1988), p.64. It may seem strange to introduce Aristotle suddenly like this. We are accustomed to think of Aristotle as an outmoded philosophical system builder, whereas in fact nothing could be further from the truth. He did not in fact build a system, even though he has been presented

as doing just that. This is a misunderstanding that arose from the way that Aristotle's work was (mis)interpreted in the Middle Ages. In a lecture given in Oxford in 1986 on 'Plato and Aristotle in the Thought of Heidegger' (British Society for Phenomenology Conference on 'Phenomenology and Ancient Greek Philosophy'), Gadamer said, referring to Aristotle, that 'he did not have ultimate foundations; there is no first principle – all this was added later'. He added that, for Heidegger, Aristotle was a phenomenologist because he had no system.

It is especially within the phenomenological tradition that Aristotle's philosophy has come to life again, and particularly through Heidegger. It is now recognised that Heidegger's concern with Aristotle deeply influenced his own thinking in his 'phenomenological decade' leading up to the publication of *Being and Time* in 1927. Many who attended his lectures have attested to the extraordinary impact they had at the time, especially his reading of the Greeks, and Aristotle in particular:

> Most striking, then, was Heidegger's ability to see Aristotle not as a historically important object, but as a way of clarifying the most pressing and urgent question of the time, namely, that of life. In his writings, and more so even in his teaching, Heidegger was able to make the Greeks speak *as if for the first time* by anchoring their thought in the fundamental experiences of human existence. His phenomenological and hermeneutical approach brought the canonical Greek texts back to life by bringing them back into the concrete life-world of our own experience (the "factical life") ... Heidegger's students had the impression that the Greeks were speaking to them directly across the ages, and that the questions of the Greeks were – or had become – their own. This, in effect, was the source of what Gadamer [who was there] called the 'fundamental hermeneutic experience', which became the focus of his own philosophy. (Miguel de Beistegui, *The New Heidegger*, pp.191–92.)

My own experience of Aristotle coming to life through Heidegger, happened to me quite unexpectedly one morning. At the time I was living with my wife in a small cottage on the outskirts of Harare in Zimbabwe. On the morning in question, I was sitting in an armchair on the veranda reading Charles Guignon's book *Heidegger and the Problem of Knowledge*. At the bottom of page 98, I read 'there is no way to drive a wedge between an "I" and the world to which it is related', which was followed immediately by a quotation from Heidegger's lecture course *The Basic Problems of Phenomenology*: 'Self and world are not two entities, like "subject" and "object" ... rather self and world are the basic determination of Dasein itself in the unity of the structure of Being-in-the-world'. As I read this it seemed as if the page in front of me 'opened', and I had the sense of Aristotle's concrete philosophy of the event coming towards me from within the text. I have no idea why it happened to me like this – perhaps it was the heat and the sounds of Africa surrounding me. What I experienced in this dynamic way was the intuition of Aristotle's account of concrete activity coming to life again in Heidegger's phenomenology of everydayness – especially in the unitary structure of being-in-the-world. There was an overwhelming sense of the past coming to life in a living philosophical tradition which is transformational and anything but conservative.

10. Lear, *op. cit.* p.60. I am following Lear's account throughout this section because it is the most illuminating guide to Aristotle's philosophy that I have come across. I am not alone in this. The *Times Higher Educational Supplement* said: 'Whether one is approaching Aristotle for the first or the *n*th time, it is hard to think of a more enlightening or engaging companion than Jonathan Lear'.

11. Ibid.

12. *Ibid.* p.31.

13. *Ibid.* p.32/3. It is not only in science but also in modern philosophy that we have become accustomed to separating cause and effect into two events. We see this very clearly in Hume's description of causality, for example, where it is a consequence of this separation into one event (effect) following another (cause) that he is led to conclude that we cannot observe the actual *causing*. It is a surprising, but inevitable consequence of the two-event description that the *causing* becomes unobservable – which is quite a shock for an empirically based science to discover. Aristotle, by contrast, does not have this problem because 'while for Hume causation must be understood in terms of a relation between two events, for Aristotle there is only one event – a change' (Lear, p.31). It is interesting that Goethe's conception of causality is in line with Aristotle's: 'cause and effect should not be separated – the two together constitute the indivisible phenomenon'.

14. Lear, p.32. The *Physics* is the foundational text for understanding Aristotle because it is here that he introduces the single actualisation description of change, which then forms the basis for his account of perception and understanding in *On the Soul (De Anima)*. In the lecture already referred to (in note 9), Gadamer emphasised that the *Physics* is central for Aristotle, compared with which he said the *Metaphysics* is 'a marginal extension' – which is the opposite of the traditional view. The single actualisation

description of change is not only closer to experience, but also seems to be reflected in the structure of the Greek language itself – see, for example, Lear's account of the single activity of sounding/hearing (p.107–8). In the same lecture, Gadamer said that 'concepts are guidelines – lines guided by language'.

15. Gadamer, *Truth and Method*, p.167 and p.164.

16. *Ibid.* p.164, and Richard J. Bernstein, *Beyond Objectivism and Relativism: Science, Hermeneutics, and Praxis* (Oxford: Blackwell, 1983), p.139.

17. Gadamer, *op. cit.* p.164.

18. Jason Elliot, *An Unexpected Light: Travels in Afghanistan* (London: Picador, 1999), pp.408–9.

19. In an interview, on the occasion of his hundredth birthday, which appeared in the *Frankfurter Rundschau*. See Robert J. Dostal, 'Gadamer: the Man and his Work', in Robert J. Dostal, ed., *The Cambridge Companion to Gadamer* (Cambridge: Cambridge University press), 2002), p.29. It is surprising how often Gadamer's statement (*Truth and Method*, p.474) is misinterpreted to imply that he is identifying being with language. It would perhaps be clearer if a comma were to be added: 'Being that can be understood, is language'. This is clearly *not* saying that all being is language.

20. Quoted in Alan P. Cottrell, 'The Resurrection of Thinking and the Redemption of Faust: Goethe's New Scientific Attitude', in Seamon and Zajonc, *Goethe's Way of Science*, p.267.

21. In *History of the Concept of Time: Prolegomena* (Bloomington: Indiana University Press, 1985), Heidegger says '*Intentio* literally means *directing-itself-towards*' (p.29).

22. Palmer, *Hermeneutics*, p.128.

23. Gadamer, *Truth and Method*, p.268.

24. Gerald L. Bruns, *Hermeneutics Ancient and Modern*, (New Haven: Yale University Press, 1992), p.146. Chapter 7 on 'Luther, Modernity, and the Foundation of Philosophical Hermeneutics' is particularly illuminating on the hermeneutic reversal.

25. Gadamer, *op. cit.* p.95.

26. Bernstein, *op. cit.* p.139; also Gadamer, *op. cit.* p.164.

27. Gadamer, *op. cit.* p.118.

28. Joel C. Weinsheimer, *Gadamer's Hermeneutics: A Reading of 'Truth and Method'* (New Haven: Yale University Press, 1985), p.100.

29. Gadamer, *Truth and Method*, p.115.

30. Weinsheimer, *Gadamer's Hermeneutics*, p.110.

31. Weinsheimer, *op. cit.* p.111.

32. Wachterhauser, *Beyond Being,* p.157.

33. *Ibid.* p.146. Hans-Georg Gadamer, *Hegel's Dialectic: Five Hermeneutical Studies* (New Haven: Yale University Press, 1976), p.94.

34. Jean Grondin, *Sources of Hermeneutics* (Albany: State University of New York Press, 1995), p.x.

35. Brice R. Wachterhauser, 'History and Language in Understanding', in Brice R. Wachterhauser, ed., *Hermeneutics and Modern Philosophy* (Albany: State University of New York Press, 1986), p.34. The expression 'circle of the unexpressed', which Gadamer frequently quoted, is due to Hans Lipps, *Untersuchungen ze einer hermeneutischen Logik,* Frankfurt: Klostermann, 1938, p.71). See also p.xxxii of the work referred to in note 38 below.

36. Gadamer, quoted in Grondin, *op. cit.* p.94. The essay where Gadamer says this, 'What is Truth?' (1957), is reproduced in Brice R. Wachterhauser, ed., *Hermeneutics and Truth* (Evanston: Northwestern University Press, 1994), pp.33–46 (see p.42).

37. Gadamer, *Truth and Method,* p.458. It should perhaps be emphasised that when an author is trying to bring something into expression, what he is concerned with is the subject matter itself, and not with expressing himself. So when we are trying to understand what the author says, we are not trying to understand the author, but what he says about the subject matter. Anyone who believes that he or she is trying to understand the author is simply on the wrong track. So, for example, I am the author of a book on the philosophy of Goethe's way of science, and anyone reading that book will want to understand what it is trying to express *about Goethe's way of science.* The intention is to say something about this subject matter, not to provide a means for the author to express himself. Anyone who reads the book will be trying to understand what it says about the subject matter. They will not – unless they are very confused – be trying to understand the author. It was the error of nineteenth century Romanticism, with its emphasis on individuality and genius, which invited the confusion of thinking that 'expression' meant 'self-expression'. Although we have now for the most part succeeded in freeing ourselves from the limitation of such subjectivism, and in the process of doing so come to understand the illusion on which it is based, it does nevertheless still catch us out again from time to time. Of course, we often do talk in a shorthand way about trying to understand Plato, for example, when what we mean is that we are trying to understand the subject matter, i.e. what it is that Plato is endeavouring to bring to expression.

38. David E. Linge, in the Introduction to Hans-Georg Gadamer, *Philosophical Hermeneutics* (Berkeley: University of California Press, 1977), p.xxvi.

39. Gadamer, *Truth and Method,* p.118.
40. Weinsheimer, *Gadamer's Hermeneutics,* p.111.
41. Gadamer, *Truth and Method,* p.148.
42. *Ibid.* p.373 and p.398.
43. Chapter 3, note 17.
44. Joel Weinsheimer, *Philosophical Hermeneutics and Literary Theory* (NewHaven: Yale University Press, 1991), p.19.
45. See the editor's introduction in Lawrence K. Schmidt, ed., *The Specter of Relativism: Truth, Dialogue, and 'Phronesis' in Philosophical Hermeneutics* (Evanston: Northwestern University Press, 1995)
46. Wachterhauser, *Beyond Being,* p.143. Wachterhauser points out that 'Zuwachs' often means growth, and so 'its literal meaning connects the idea of an "increase of being" to a natural organic process' (p.153). We must guard against thinking of understanding meaning as the actualising of a potential in the manner of physics – in which case the meaning actualised in understanding would become finished. The work's potency to mean is more like living being, a condition of existence, and hence understanding is not like a process that finishes when it reaches an end result. On the contrary, the work goes on meaning. As Agamben says: 'Contrary to the traditional idea of potentiality that is annulled in actuality, here we are confronted with a potentiality that conserves and saves itself in actuality' – see Giorgio Agamben, *Potentialities: Collected Essays in Philosophy* (Stanford: Stanford University Press, 1999), p.184. Risser makes the point that 'the ontology of a philosophical hermeneutics can also be said to be an ontology of living being', and goes on to say: 'what is required by an ontology of living being (is) a reversal in the priority between possibility and actuality so that possibility stands higher than actuality' – see James Risser, *Hermeneutics and the Voice of the Other: Re-reading Gadamer's Philosophical Hermeneutics* (Albany: State University of New York Press, 1997), p.124. What this means is that understanding is not the actualisation of a potential as a process of unfolding, as if the meaning were a potential waiting to be realised, because what is there for understanding is a possibility that *always remains as possibility.* This is what we find when we think in a manner that is more in accordance with life.
47. I was first introduced to the idea of 'increase in being' in a very different context. In the 1960s I was a student of the mathematician and philosopher, J.G. Bennett, who developed a philosophy of time based on the Minkowski representation of the four-dimensional space-time of Special Relativity. Bennett extended this framework to include three dimensions of time

instead of the usual one. He and his colleagues had found this led to a richer geometry of null vectors (the null interval of the light cone in the Minkowski diagram) than was possible otherwise, which provided a more adequate framework for developing a unified field theory (at the time, this approach of increasing the number of dimensions was generally dismissed by physicists, whereas now the introduction of extra dimensions has become commonplace). Bennett was particularly interested in the wider philosophical implications of these ideas about the dimensionality of time, especially the notion of 'ableness-to-be', for which he adopted the term *hyparxis* (a term originally used by Aristotle, and later by the Neoplatonists, especially Proclus). Instead of a polar opposition between potentiality and actuality, which are mutually exclusive (because what is actual ceases to be potential, whereas what is potential has no actuality), ableness-to-be (*hyparxis*) is a third factor whereby potentiality and actuality can be reconciled in a way that allows something to become itself without losing contact with its own potential. An essential feature of *hyparxis* is that, unlike space and time, it is discrete instead of continuous. Its major characteristic is return or recurrence, which Bennett emphasised is not the same as repeated actualisation in time. He identified meaning specifically with *hyparxis*, and we have seen that the return of the same, differently, is a fundamental feature of meaning in relation to understanding a text, or presenting a play, or a piece of music, and so on. See J.G. Bennett, *The Dramatic Universe Volume 1: The Foundations of Natural Philosophy* (London: Hodder and Stoughton, 1956), p.135 and pp.166–70.

What Gadamer refers to as the enhancement or 'increase in being' which results from the multiple interpretation of a work, seems to correspond to what Bennett referred to as 'progress in hyparchic depth'. The work increases in being because its meaning becomes more fully itself (i.e. able-to-be) through the multiple occasions of its presentation in different contexts and situations. This is not a process in time in the usual sense because it is always the very same One, but differently, and not another one – which it would be if it were just a case of repeated actualisations in time, because then there would be many *Hamlets* instead of One *Hamlet* manifesting multiply. So this is a different *kind* of temporality and not just repetition *in* time in the usual sense. If for the moment we adopt Bennett's terminology, we could say that the 'multiplicity in unity' (forming an intensive dimension of One) which results from the dynamic unity of self-differencing, is the *hyparchic unity* of the work and its interpretations. An art work cannot be detached from its presentations, so that it *is* the work itself which comes-to-presence, i.e. *self-presentation* is characteristic of the art work. This self-presentation is discrete because it is clearly occasional, and the unity of these occasions of its self-presentation is hyparchic because it is the work itself which recurs, even though it will do so differently on each occasion. Gadamer says that:

> Inescapably, the presentation has the character of a repetition of the same. Here "repetition" does not mean that something is literally repeated – i.e. can be reduced to something original. Rather, every repetition is as original as the work itself. (*Truth and Method*, p.122.)

He says that this gives the work of art a 'contemporaneity', which 'means that in its presentation this particular thing that presents itself to us achieves full presence, however remote its origin may be' (*Ibid.* p.127)

Gadamer goes on to give the celebration of a festival as an example. Each time a festival is celebrated it is neither a new festival nor a remembrance of an earlier one. In celebrating a festival we are not simply repeating the first festival. The festival which we celebrate is that festival now, not just a repetition but contemporaneously – i.e. it is not just a matter of the survival of something from the past, but of something coming to life again now. Speaking about this return of the festival, Gadamer says: 'But the festival that comes round again is neither another festival nor a mere remembrance of the one that was originally celebrated' (*Ibid.* p.123), and he then goes on to say: 'The temporal character of celebration is difficult to grasp on the basis of the usual experience of temporal succession'. In Bennett's philosophy of time this would be because the temporal character of celebration is hyparchic and not repeated actualisation in time. From the perspective of the dynamical mode of thinking discussed here, we would say that this takes the form of the dynamic unity of self-difference, and hence 'multiplicity in unity', i.e. we have to think of celebrating a festival in the intensive dimension of One and not the extensive dimension of many ones. This is where the unity of the festival is to be found.

48. Darwin describes his own experience of the variation in domesticated pigeons (as well as other organisms) in the first chapter of *The Origin of Species*.

49. Wachterhauser, *Beyond Being*, p.149.

50. Gadamer, *Truth and Method*, p.312.

51. Gadamer goes further, and sees the same reciprocal movement between universal and particular in the kind of practical reason that Aristotle calls *phronesis*. The importance of this is that Gadamer is able to show how it

applies in our own situation today, and consequently how our concrete situation enables us to understand what Aristotle is saying about this in the *Nicomachean Ethics*. Here we see an illustration of the way that Gadamer's philosophical hermeneutics provides a concrete instance of the very hermeneutical situation with which it is itself concerned. In this case it is not a matter of bringing something back from the past, but of it coming to life again now, coming-into-being in a situation which is very different from the original one.

We cannot go into details here, but suffice it to say that this kind of practical knowledge – which is different from either theoretical knowledge of universal principles (*episteme*) or knowledge of techniques (*techne*) – which concerns choosing the right action in a concrete situation, has the same mutual codetermination of the universal and the particular that we have seen already. In this case 'The choice that is right cannot be determined in advance or apart from the particular situation, for the situation itself partly determines what is right' (Weinsheimer, *Gadamer's Hermeneutics*, p.190). This is in contrast to those who seek for ethical universals in the mathematical sense, i.e. moral principles which are invariant, the same for everyone at all times and under all circumstances. Such an approach takes no account of the individual case because everything is subsumed under the universal, of which the particular is merely an instance. There certainly have been many who believed in such moral imperatives, which abstract from the specifics of concrete cases to search for 'the good' in the form of a timeless universal principle. But there have been others who thought differently, believing that such an abstract notion of the good is of little use in practice: 'Ethical knowledge is not knowledge that a specialist or theoretician can discover for others once and for all; it is not the same as a theory of the good or an account of a separate and unchanging universal, a charge Aristotle sometimes levels at Plato' (Georgia Warnke, 'Hermeneutics, Ethics, and Politics', in Robert J. Dostal, ed., *The Cambridge Companion to Gadamer*, p.82). Aristotle opposed this kind of universalisation, emphasising instead that the good has no universal form regardless of the situation, and that the judgement of what is the right thing to do must always take the specific circumstances into account.

52. Gadamer, *Truth and Method*, p.309. The translation given here differs slightly from that given in the English translation of *Truth and Method* by the addition of the word 'already', which helps to make the point more clearly. See James Risser, *Hermeneutics and the Voice of the Other*, p.102.

53. Gadamer, *Truth and Method*, p.341 and p.324.

54. Weinsheimer, *Gadamer's Hermeneutics*, p.185.

55. Gadamer, p.308; Weinsheimer, p.185.

56. Gadamer, *Ibid.* p.391 and p.392.

57. *Ibid.* p.332.

58. Gadamer refers to 'the paradox that is true of all traditionary material, namely of being one and the same and yet of being different', and he finds an analogue of his own thought about the same and the different in Plato's idea of the one and the many (Gadamer, *Truth and Method,* p.437; Weinsheimer, *Gadamer's Hermeneutics,* p.256). What we have found here is an analogue with the idea of 'the one and the many' as the dynamic unity of self-differencing that we discover in life. In this respect, we can refer to Gadamer's approach as 'organic' hermeneutics. Scheibler points out that, for Gadamer, 'continuity is an identity *constituted by difference*' – see Ingrid Scheibler, *Gadamer: Between Heidegger and Habermas* (Lanham: Rowman and Littlefield, 2000), p.148. We can easily miss this, because we tend to think of continuity as constituted by sameness – an identity constituted by the persistence of the same throughout difference. Brady finds the same confusion in the way we see the dynamics of the plant. In the context of comparing T.H. Huxley with Goethe, he says:

> We can take the continuity of the series as an indication of a common underlying schema only by a sort of mental laziness – we do not care to undertake the problem of how things may be united by *difference*, preferring the empty alternative that they were not really different at all – that is, they are united by sameness. (Ron Brady, in Amrine, Zucker and Wheeler, *Goethe and the Sciences: A Reappraisal,* p.277.)

Chapter 5

1. Gadamer, *Hegel's Dialectic,* p.94.

2. David Mitchell, *An Introduction to Logic* (London: Hutchinson, 1962), p.101.

3. Maurice Merleau-Ponty, *Phenomenology of Perception,* p.177 and p.178.

4. Weinsheimer, *Gadamer's Hermeneutics,* p.228.

5. Mitchell, *An Introduction to Logic,* p.101: 'Wordsworth is reported to have said that language is not the clothing but the incarnation of thought'.

6. Charles Taylor, *Sources of the Self: The Making of the Modern Identity* (Cambridge: Cambridge University Press, 1989), p 374. See also Wachterhauser, *Beyond Being,* p.145.

7. Wachterhauser, op.cit. p.145.

8. *Ibid.* p.96.

9. Mitchell, *An Introduction to Logic*, p.101.

10. Charles Taylor, 'Heidegger on Language', in Hubert L. Dreyfus and Mark A Wrathall, eds., *A Companion to Heidegger*, p.439 and p.437.

11. Helen Keller, *The Story of My Life* (London: Hodder and Stoughton, 1959), p.23.

12. Martin Heidegger, *On the Way to Language* (New York: Harper and Row, 1971), p.123 (Heidegger's italics) and p.126. The three quotations that follow are all taken from p.115.

13. Bryan Magee, *Confessions of a Philosopher* (London: Weidenfeld and Nicholson, 1997), p.96. The relevant part of the interview with A.J. Ayer is on p.93.

14. *Ibid.* The quotation is on p.96, and the example of the potato is on p.97.

15. Heidegger, *On the Way to Language,* p.119.

16. Risser, *Hermeneutics and the Voice of the Other,* p.134 and p.149.

17. Günter Figal, 'The Doing of the Thing Itself', in Dostal, ed., *The Cambridge Companion to Gadamer,* p.115.

18. Gadamer, *Truth and Method,* p.474.

19. Michael E. Zimmerman, *Eclipse of the Self: The Development of Heidegger's Concept of Authenticity* (Athens: Ohio University Press, 1986), p.233. Zimmerman emphasises the fundamental difference between Heidegger's notion of 'Being' and the way that we usually understand this in the English language:

> Being is a translation of the German *Sein,* a substantive form of the infinitive *sein,* meaning "to be". Our word Being is a gerund which lacks the fully active sense of the infinitive, to be. Hence, when we hear the word Being we often think of something which *is,* a "being" *(Seiende).* Heidegger stresses, however, that Being does not mean any particular being, or even the totality of beings. Being ("to be") means for a being to be manifest or revealed. Being, refers, therefore, to a happening or an event. We must hear the infinitive "to be" whenever we read the word Being in translations of Heidegger. (p.xxxii)

20. Charles E. Scott, *The Lives of Things* (Bloomington: Indiana University Press, 2002), p.62. See also Michael E. Zimmerman, *Heidegger's Confrontation with Modernity: Technology, Politics, Art,* (Bloomington: Indiana University Press, 1990), p.167 and p.228.

21. Terry Eagleton, *How to Read a Poem* (Oxford: Blackwell, 2007), p.67.

22. *Ibid.* p.68.

23. Georg Kühlewind, *The Logos-Structure of the World: Language as Model of Reality* (Hudson, N.Y.: Lindisfarne Press, 1991), p.31.

24. Gadamer, *Truth and Method,* p.417.

25. Ibid.

26. Patricia Altenbernd Johnson, *On Gadamer* (Belmont: Wadsworth, 2000), p.45.

27. Kuhlewind, *op. cit.* p.52.

28. In Chapter 3, we noted the emphasis Schelling placed on the need to rise from nature as fact to nature as 'the action itself in its acting' – from *natura naturata* ('nature natured') to *natura naturans* ('nature naturing') – and that he said: 'In the usual view, the original productivity of nature disappears behind the product. For us the product must disappear behind the productivity.' Wilhelm von Humboldt emphasised that language 'In itself is no product (*Ergon*), but an activity (*Energeia*)', and so 'We must look upon language, not as a dead *product*, but far more as a *producing*'. (Wilhelm von Humboldt, *On Language: On the Diversity of Human Language Construction and its Influence on the Mental Development of the Human Species,* Cambridge University Press, 1999, p.49 and p.48.) Thus Humboldt's dynamic approach to language is entirely consistent with Schelling's dynamic approach to nature – which, as we have seen, is very much that of Goethe. The similarity is not surprising because they were all in the cultural milieu within which the dynamical idea emerged at the end of the eighteenth and beginning of the nineteenth centuries.

29. Gilles Deleuze, *Bergsonism,* p.42. See Chapter 3, note 9 above.

30. Mark Abley, *Spoken Here: Travels Among Threatened Languages* (London: William Heinemann, 2003), p.196.

31. Quoted by James Edie in his foreward to Maurice Merleau-Ponty, *Consciousness and the Acquisition of Language* (Evanston: Northwestern University Press, 1973), p.xxix. See Maurice Merleau-Ponty, *The Prose of the World,* (Evanston: Northwestern University Press, 1973), p.39 (translation altered slightly by Edie).

32. Weinsheimer, *Gadamer's Hermeneutics,* p.243. When language is considered as an object in itself apart from what is said in it, i.e. apart from what it means, then the focus is on grammar, syntax, etc., as if these were primary structural elements from which the language is built in the first place, instead of recognising these as secondary abstractions from living language. Aristotle's philosophy of mathematics may provide us with a useful analogy here. According to Aristotle, mathematics is not about a separate realm of pure mathematical objects, but is a way of thinking about objects that exist in the physical world. What mathematics does is to think about properties of physical objects in a special way, namely in abstraction from

their physical instantiation. The geometer studies lengths, areas, volumes, etc., but in isolation from their physical instantiation. In separating them in thought from their instantiation, he is still studying physical bodies but not *as* physical. Aristotle says that in doing so it is convenient to suppose that there are mathematical objects that satisfy these geometrical properties alone. This is a fiction, but it is a helpful one. It goes wrong when we forget that it *is* a fiction, and believe that there is actually a realm of mathematical objects separate from physical bodies. When this happens, methodology is mistaken for ontology (Aristotle's philosophy of mathematics is discussed very clearly in Jonathan Lear, *Aristotle*, pp.231–47). It seems that something similar to this is also the case with the science of linguistics, where 'language' is abstracted from what is said in it and considered as if it existed as such – i.e. as a pure form – independently of the fact that in any instance of language something is said. This fiction of language as pure form may well be useful for some purposes, but when it is forgotten that it *is* a fiction, and we believe that language exists as such independently of what is said, then we fall into the form-content dualism which leads us straight back into the instrumental theory of language.

33. Ibid.

34. Ibid.

35. Guignon, *Heidegger and the Problem of Knowledge*, p.191. See also Simon Critchley, *Continental Philosophy: A Very Short Introduction* (Oxford University Press, 2001), p.104.

36. Gadamer, *Truth and Method*, p.450.

37. Heidegger's shift from the epistemological approach – which is based on the subject-object model of our perception of physical objects – to the notion of 'world' in the hermeneutic phenomenology of lived experience, is described very clearly in Cristina Lafont, 'Hermeneutics', Dreyfus and Wrathall, *A Companion to Heidegger*, pp.265–284.

38. Timothy Clark, *Martin Heidegger* (London: Routledge, 2002), p.16. In note 21 of the first chapter, we saw how, in his first lecture course (1919), Heidegger tried to make his students aware of what they directly encounter in lived experience. The quotation we gave there will bear repeating here:

> This environmental milieu (*Umwelt*) ... does not consist just of things, objects, which are then conceived as meaning this and that; rather, the meaningful is primary and immediately given to me without any mental detours across thing-oriented apprehension.

He immediately goes on to say:

> Living in an environment, it signifies to me everywhere and always, everything has the character of world. It is everywhere the case that 'it worlds' (es weltet)...

Michael Watts comments on this:

> Heidegger's innovative "it is worlding" is used in an attempt to convey something for which no single expression exists, because what is being described normally goes unrecognised, first, because it is too close to us, and second, because our habitual conditioned patterns of response tend to block out such experiences from conscious awareness. "Worlding" is a word Heidegger uses to refer to the *dynamic presence* of the world. Our primal experience and interpretation of ourselves and our surrounding environment is always as a living totality [whole], and not in terms of individual characteristics or entities in isolation, and the "activity" contained in the word "worlding" expresses the energetic aliveness – the *presencing* of an environment that is a *process* in constant flux ... (Michael Watts, *The Philosophy of Heidegger*, p.66.)

An illuminating discussion of 'worlding' is given in Rüdiger Safranski, *Martin Heidegger: Between Good and Evil* (Cambridge MA: Harvard University Press, 1998), pp.94–97.

39. Weinsheimer, *Gadamer's Hermeneutics*, p.230.
40. Gadamer, *Truth and Method*, p.443.
41. Ibid.
42. *Ibid.* p.450.
43. *Ibid.* p.447. I have modified the quotation slightly here by putting 'view' where Gadamer puts 'views'. The reason for this small change is apparent from the context. We have been considering the internal relationship between language and world, whereas Gadamer is referring here to the multiplicity of worldviews which emerge with the diversity of language. We will be going on to consider this below.
44. Charles Guignon, *Heidegger and the Problem of Knowledge*, p.125.
45. Eagleton, *How to Read a Poem*, p.68. We have noted previously that poetry negates this transparency, so that in seeing the meaning take shape our attention is at the same time drawn to the words themselves – so 'it allows us to experience the very medium of our experience'. (*Ibid.*)

46. If we did, language would have to be a private experience. But if language were a private experience, we would each first have a private world, and then each of these separate, private worlds would somehow have to be related to produce the public world. But this could only come about if the public language to which it is supposed to lead already existed. Although there have been philosophers – Thomas Hobbes, for example – who believed that this is indeed how the public world is formed, in fact this gets it back to front. The philosophy of language, which became such a dominant theme in the twentieth century, shows this clearly – major philosophers from different philosophical traditions, such as Heidegger and Wittgenstein, are agreed on this. This shows us a way out of the cul-de-sac into which we are led by the Cartesian project that tries to *begin* with the self-certainty of I-consciousness. Heidegger took one route out of this cul-de-sac, while Wittgenstein took another – although there is more overlap between them than is sometimes recognised. Heidegger's way of overcoming the 'problem' of epistemology (he sees it as a pseudo-problem) is presented in *Being and Time*. A clear account of this aspect of Heidegger's philosophy is given in Charles Guignon, *Heidegger and the Problem of Knowledge*. Wittgenstein's way beyond the Cartesian illusion of what is often referred to as 'first-person certainty', is to be found in his later philosophy of language, especially in those parts of his *Philosophical Investigations* dealing with what has come to be known as the private language argument. Although this is not always easy to follow, it is nevertheless recognised as being of fundamental importance in showing the way out of the Cartesian impasse. See Roger Scruton, *A Short History of Modern Philosophy: From Descartes to Wittgenstein,* second edition (London: Routledge, 1995), p.281 *seq.* See also the same author's *Modern Philosophy: A Survey* (London: Sinclair-Stevenson, 1994), Chapter 5.

47. For most of us this is the dominant aspect of our experience in adult life. But during the last few decades, it has become clearer that this 'higher level' activity emerges from a prior level of embodied image-schemata which are built up through our experience as embodied beings in a world of bodies. There is a hitherto neglected level of imagination prior to linguistic meaning, and upon which the latter depends. Some of the functions which we think of as being 'higher' – the ability to reason and think logically – turn out to be rooted in such embodied image schemata. But the explicit articulation in which this comes into expression and thus *appears* depends on the disclosure of language. See Mark Johnson, *The Body in the Mind: The Bodily Basis of Meaning, Imagination, and Reason* (Chicago: Chicago University Press, 1987)

48. Gadamer, *Truth and Method,* p.450 and p.447 (author's italics)

49. See, for example, Franz Meyer, 'The Question of Being, Language, and Translation', in Martin Heidegger, *Zollikon Seminars: Protocols – Conversations – Letters,* edited by Menard Boss (Evanston: Northwestern University Press, 2001), pp. 317–336. Meyer is one of the translators of this work, and in this afterword he reflects on the way that linguistic differences between the German and English languages come to expression in differences between their respective philosophies. For example:

> While the German language is more holistic, historically oriented, synoptic, and interconnective, the English language is given to pluralism, favours the concrete, the empirical, the particular, and the "given". It takes a nominalistic approach to reality. The term "reality" is from the Latin *res* (thing). In German it is *Wirklichkeit,* from *wirken* (to be active or effective). This term implies action, activity, and an orientation to the future. (p.319)

> In its thought and speech patterns, German is more contextual than English and more synthetic than analytic. (p.320)

> In German understanding, language is primarily "expressive", concerned with the internal unity of meaning, feeling, and contextual reference ... For the English speaker, language is predominantly an instrument based on the conventional, "representative" sign character of the language which is similar to the Latin relationship between *res* and *signum.* It is interesting to note that a similar nominalistic understanding of language underlies much of the philosophy of language in the English-speaking world. (p.320)

> The English language has an atomistic view of being, which tends to reduce being to discrete entities and objects. This view underlies modern logic, mathematics, and science. (p.321)

> In a manner similar to its "atomistic" understanding of being, English also prefers contingent "external relations" between beings (entities), which can be formalized logically ... In contrast, the German language has a preference for understanding and expressing "internal relations", that is, the immanent interconnections of things with things and the relations of part to whole or of whole to part. German vocabulary is embedded in historical context, social relationship, and interaction, that is, the internal constitutive relationship which constitutes the nature of things, persons, and events. (p.323)

215

From the time of Goethe onwards, the Germans have been aware of an important difference between Greek and Latin, and the distortion that Greek thinking underwent when it was translated into Latin. In the case of Goethe:

> In different languages he identified different kinds of expressivity; ancient Greek for example, is a dynamic language because of its abundance of verb forms and verbal nouns and adjectives, whereas the nominative Latin reifies and abstracts so that its overall form is static and monumental. (Dennis L. Sepper, *Goethe contra Newton: Polemics and the project for a new science of colour*, p.93.)

Emilios Bouratinos told me (in conversation at a conference on 'The Evolution of Consciousness' at the Institute of Psychiatry, University of London, November 1999) that he thought it was a tragedy that the West had received Greek filtered through the prism of Latin. Overall the influence of Latin on the development of the western mind has been far greater than Greek – especially in view of the way that the 'Latinisation' of Greek replaces the Greek emphasis on the sensuous immediacy of experience with the Latin emphasis on abstract concepts. It is possible that many of the ideas we attribute to Greek philosophy may be distorted because they have been passed through the 'prism of Latin'. The two-world dualism of 'Platonism' – with its ontological *separation* between a changing world of appearances and an immutable world of being-as-it-is-in-itself – may be a consequence of this process of transmission. Such a static notion of being – from which the very possibility of difference is excluded – seems more in tune with the reifying and abstracting tendency of Latin than with the dynamic verbal character of Greek.

50. Steven Levinson, of the Max Planck Institute of Psycholinguistics in the Netherlands, *The Independent*, Friday, 25 January, 2008. See Abley, *Spoken Here* (note 30), and K. David Harrison, *When Languages Die: The Extinction of the World's Languages and the Erosion of Human Knowledge* (New York: Oxford University Press, 2007).

51. Edward Sapir, quoted in Abley, p.47.

52. F. David Peat, *Blackfoot Physics: A Journey into the Native American Universe* (London: Fourth Estate, 1996), p.222, and p.220.

53. Leibniz, quoted in Stephen Toulmin, *Return to Reason*, (Cambridge MA: Harvard University Press, 2001), p.70.

54. Stephen Toulmin, *Cosmopolis: The Hidden Agenda of Modernity*, (Chicago: University of Chicago Press, 1992), p.103.

55. The group came to a focus around Moritz Schlick. Prominent members were Hans Hahn, Otto Neurath, and Rudolf Carnap. It is sometimes said that Wittgenstein was a member of the Vienna Circle. But this is not really correct. Wittgenstein had an enormous influence on them, especially through the 'logical atomism' of his *Tractatus Logico-Philosophicus*. However, by the time the Vienna Circle got going (1929), Wittgenstein had moved beyond his earlier ideas and was reluctant to engage himself with them. In fact he had come to think that those aspects of his earlier work which were so attractive to the Vienna Circle were fundamentally flawed. He rejected the idea of logical atomism because he came to realise that there are internal logical connections between propositions. But more importantly for our interest here was the radical change in his attitude to language. He ceased to think that language was in need of improvement by means of the new developments in mathematical (symbolic) logic, and instead came to think that philosophers should learn how to work with natural languages just as they are. It is ironic that, at the very time the Vienna Circle was hoping he would come to their meetings (he was living again in Vienna at the time), he was moving towards an understanding of language as a 'form of life' which was in many ways diametrically opposed to their approach. Karl Popper was also in Vienna at that time, and although he always distanced himself from the Vienna Circle, we can now see that his overall approach was very much in sympathy with them in spirit, whilst at the same time differing from them in important respects – so much so that Otto Neurath nicknamed him 'the Official Opposition'. It could be argued that for a long time Popper's philosophy was not properly understood because it was always in the shadow of 'logical positivism' – this being the name generally given to the philosophical doctrine of the Vienna Circle ('positivism' from the French *positif*, meaning 'sure', 'certain'; the alternative name, 'logical empiricism', is a better one, but not so widely used). The young A.J. Ayer spent several months in Vienna in 1932–33, as a result of which he wrote his highly readable and influential account of the Vienna School, *Language, Truth and Logic* (1936). So when Popper came to live in England in 1946, he found the prevailing philosophy at the time was the logical positivism he had left behind him in Vienna before the war. During the 1950s this influence gradually began to erode with the rise of linguistic philosophy stemming from the later Wittgenstein (of which Popper also did not approve). But in America the influence of the Vienna Circle and logical positivism lasted a lot longer. Many members of the Vienna Circle left after the annexation of Austria by Germany in 1936 and went to America, where they had considerable impact.

Carnap in particular 'had a vast influence on the professional development of analytical philosophy in the United States after the Second World War, not least through his most celebrated student, W.V.O. Quine' – in a eulogy given after Carnap's death in 1970, Quine 'describes philosophy in the United States after the Second Word War as "post-Carnapian", rather than "post-Wittgensteinian", which arguably describes the comparable period in Britain'. (Simon Critchley, *Continental Philosophy*, p.91.)

It is astonishing to see how these philosophical ideas, developed at a particular time and place by a small group of people, could become so widespread – although concern with the problem of 'the nature and limits of language, expression and communication' was already a well-established feature of Viennese culture in the nineteenth century – Allan Janik and Stephen Toulmin, *Wittgenstein's Vienna* (New York: Simon and Schuster, 1973) p.117. The idea that the logical analysis of language is a necessary foundation for philosophy is fundamental to logical positivism – and as we shall see, this takes us in a very different direction to the hermeneutical experience of language. We saw how, in the case of Leibniz, the idea of a universal language did not just fall out of the sky, but was motivated by the desire to overcome the religious, political, and social differences that had led to the Thirty Years War in the seventeenth century. There was a similar kind of motivation with the Vienna Circle's aim of reforming language to remove the ambiguities and inconsistencies which can confuse our understanding. They had all lived through the effects of the calamitous collapse of the Austro-Hungarian Empire, and also witnessed the rise of various ideologies displacing the discredited nationalism which it was seen had led to the First World War. What was needed, they believed, was a philosophy that could save them from political and social conflict, and this is where the logical reform of language was important, because it would save people from bogus claims whilst showing them what really is true and therefore can be believed. Whatever the limitation of the outcome, the motive was certainly well intentioned.

56. Since the time of Leibniz it has been customary to divide all true propositions into two classes, which he called 'truths of reason' and 'truths of fact', and which in modern philosophy (following Kant's terminology) are usually referred to as 'analytic' and 'synthetic' propositions. An analytic proposition is one which is intrinsically true because the predicate is the defining characteristic of the subject – for instance, 'all triangles are three-sided' – so that any counter-instance would be self-contradictory, which is another way of saying that there cannot be a counter-instance. For instance, in the

above example, a counter-instance would take the form 'there is a particular triangle which isn't three-sided', which is equivalent to saying 'there is a three-sided figure which isn't three-sided', which is self-contradictory. As well as all definitions, the propositions of logic and mathematics are analytic propositions. So the truth of mathematics is self-contained, and hence can be ascertained without going outside of the mathematical system itself. A synthetic proposition, on the other hand, is one that refers to something which is not contained in itself, and to which reference must be made to ascertain whether or not it is true. For example, 'There is at least one raven in Iceland' requires someone to go to Iceland and look. The counter-instance, that there are no ravens in Iceland, is certainly not self-contradictory. On the contrary, it is just as possible. Analytic propositions must be true, but synthetic propositions are empirical and so just happen to be true, and could just as well be false. So the truth of the former is necessary, whereas that of the latter is contingent. As the name 'logical empiricism' indicates, this brings together these two kinds of proposition, and it is a key feature of this philosophy that these are the *only* kinds of propositions that have meaning. Thus, according to logical positivism, for a proposition to be meaningful it must be either analytic or synthetic – there is no other possibility (this dichotomy is often known as 'Hume's fork', because he had much earlier insisted upon this as the criterion for distinguishing what is genuine knowledge from what is not). By insisting on this, the logical positivists were explicitly rejecting the still influential philosophy of Kant, which maintained that there is a third kind of proposition, one which paradoxically combines the characteristics of both analytic and synthetic propositions in one and the same proposition. These are propositions which are synthetic and yet *a priori* (so like analytic) instead of *a posteriori* (so not empirical). As it turns out, the logical positivist's division into two and only two kinds of meaningful propositions is hopelessly inadequate for understanding science, and ironically something more akin to Kant's approach (though not the same) is needed. There is a further class of propositions – which could be called 'paradigmatic' or 'constitutive' propositions – which is needed in order to understand the fundamental principles of science. An account of the philosophy of science which compares the approach taken by logical positivism with that of what is often called 'the new philosophy of science' is given in Harold I. Brown, *Perception, Theory and Commitment: The New Philosophy of Science* (Chicago: University of Chicago Press, 1977). It is ironic that the seminal work which undermined the logical positivist philosophy of science, Thomas Kuhn's *The Structure of Scientific Revolutions*, was first published (1962) in the *International Encyclopaedia of Unified Science,* which was intended to be the culmination of the Vienna Circle's programme. Another work, published just before Kuhn's by Stephen Toulmin, *Foresight and Understanding: an enquiry into the aims of science* (London: Hutchinson, 1961), gives a clear account, based on examples from the history of science, of the fundamental role of propositions in science which do not fit into the analytic/synthetic dichotomy of logical positivism.

57. The first step in this direction seems to have been taken by George Boole in his *Mathematical Analysis of Logic* (1847) – and subsequently in his masterwork, *An Investigation of the Laws of Thought* (1854). Boole put into practice the suggestion made earlier by Leibniz – but which he did not develop – that if the laws of logical thinking could be expressed in symbolic form, it would greatly facilitate our ability to recognise correct (and incorrect) logical thinking. It seems that the inspiration for this came from the development of abstract algebra, especially in Britain, earlier in the nineteenth century. There arose out of this 'a new view of algebra as symbols and operations that could represent any objects' (Kline, *Mathematics,* p.184). It was in this spirit that Boole proposed an algebra of logic – which therefore became known as 'algebraic logic'. It was in the course of doing this that Boole developed the logic of propositions, which became known as the propositional calculus and played a key role in logical positivism.

58. Kline, *Mathematics,* p.21.

59. Although it has been highly influential in western thinking, it is not universal even within mathematics. The tendency to idolise the Greek approach to mathematics has been corrected recently by the evidence of different, but equally effective, approaches to mathematics in India, China, and Arabia – see, for example, George Gheverghese Joseph, *The Crest of the Peacock: The Non-European Roots of Mathematics,* second edition (London: Penguin Books, 2000). Given the situational embedding of Aristotle's logic in the deductive reasoning of mathematics, there is clearly no reason why this logic should be universal in the sense of applying to *all* situations. So, for example, we should not expect the principles of reasoning in jurisprudence to be the same as those used in mathematics. Stephen Toulmin has explored this difference in detail, arguing that an alternative conception of logic based on jurisprudence, rather than mathematics, would provide a much more practical procedure of reasoning in many circumstances. See Stephen Toulmin, *The Uses of Argument,* updated edition (Cambridge University Press, 2003; first published 1958)

60. Gadamer, *The Idea of the Good in Platonic-Aristotelian Philosophy,* p.17/18.

61. Kline, op.cit. p.21.

62. *Ibid.* p.17.

63. Wachterhauser, *Beyond Being,* p.82.

64. Ibid.

65. Gadamer, *The Idea of the Good,* p.17.

66. Details of the proof are given in Toulmin, *Return to Reason,* p.17. See also Richard Courant and Herbert Robbins, *What is Mathematics?: An Elementary Approach to Ideas and Methods,* second revised edition (New York: Oxford University Press, 1996; first published 1941), p.59.

67. This has been especially so at times when the kind of certainty attained in mathematics has been taken as the model of what should be aimed for in *all* forms of thinking. Historically this has usually coincided with times of extreme devastation and loss of certainty. Toulmin points out in *Cosmopolis* that these were the conditions in Europe in the seventeenth century (religious intolerance and the Thirty Years War) and at the beginning of the twentieth century (the Great War and break-up of Central Europe). At both of these times there was a strong emphasis on the mathematical-logical style of thinking as the paradigm of certainty to be aspired to wherever possible.

68. Robert Blanché, *Contemporary Science and Rationalism* (Edinburgh: Oliver and Boyd, 1968), p.60. We also notice that the notion of self-differencing requires the principle of identity to be modified.

69. Gadamer, 'What is Truth?', in Wachterhauser, ed. *Hermeneutics and Truth,* p.42. See also Grondin, *Sources of Hermeneutics,* p.94 and p.106.

70. Grondin, *op. cit.* p.x.

71. Gadamer, *Truth and Method,* p.458.

72. Gadamer, quoted in Grondin, *op. cit.* p.29.

73. Martin Heidegger, *Kant and the Problem of Metaphysics* (Bloomington: Indiana University Press, 1962), p.206. But of course this is by no means restricted to works of philosophy. Safranski recounts the occasion when the physicist, Carl Friedrich von Weizsächer, told Heidegger the Jewish anecdote about a man who perpetually sits in a tavern. When asked why he does so, he answers: 'Well, it's my wife'. 'What about your wife?' 'Oh, she talks and talks and talks ...' 'What does she talk about?' 'That she doesn't say'. When Heidegger heard this story he said, 'Yes, that's how it is'. See Rüdiger Safranski, *Martin Heidegger,* p.311.

Chapter 6

1. Alfred North Whitehead, *Process and Reality* (New York: Free Press, 1979), p.39.

2. What I am considering as Platonism, and its various developments which go under the heading of Neo-Platonism, is what has come to be known as the standard established interpretation. This is the Platonism that we all recognise as such, with its defining dualism, which has had such an impact on the development of western thinking. But, as we have noted in several places (Chapter 2, note 19; Chapter 3, note 24; Chapter 5, note 49), this does not necessarily coincide with what Plato intended. So although it is with 'Platonism' that we are concerned here, we should always bear in mind that 'Plato was no Platonist' (Gadamer)

3. In the preface to *Beyond Good and Evil.* This view is endorsed by Heidegger in *Introduction to Metaphysics,* p.111. The influence of Platonism seems to have taken a different turn in Eastern (Greek) Christianity to that taken by the Western (Latin) form. In this case, at least in the early stages, the emphasis was more on participation than two-world dualism.

4. We might very well think that in order to discover the mathematical laws according to which the material world is organised, we would first have to know the properties of matter in some detail. But, surprising as it may seem at first, the mathematical laws are independent of the intrinsic properties of the matter that they organise. This often seems strange to people who, quite understandably, tend to assume that physicists must discover the mathematical laws from an investigation of the properties of matter, as if the laws are part of the matter that they organise. But Newton's law of gravitational attraction, for example, was discovered without needing to know anything whatsoever about the properties of matter. In fact, if the intrinsic nature of matter had to be understood first, it is difficult to see how mathematical physics could have developed in the first place. This fact that the law can be discovered without needing to know the properties of matter, means that the laws can readily be conceived as being separate from the matter they act upon, and hence as existing apart from the material universe they organise. In other words, the tendency to conceive the laws as being transcendent almost seems to be an inevitable consequence of the mathematical form which they take. See John D. Barrow, *The World Within the World* (Oxford: Oxford University Press, 1988), pp.35–38.

5. Margaret Wertheim, *Pythagoras' Trousers,* p.48.

6. Aron Gurwitsch, *Phenomenology and the Theory of Science* (Evanston, Illinois: Northwestern University Press, 1974), p.51.

7. Ibid.

8. Chapter 1, note 20.

9. Stephen Toulmin, *Cosmopolis*. This book is invaluable for giving the background to what the author calls 'the hidden agenda of Modernity'.

10. This was very much the aim of Descartes' project for a *mathesis universalis*, as we have seen in the first chapter. During the period following the 1914–18 war in Europe, there was a strong resurgence of the emphasis on the mathematical style of thinking as a way of reaching certainty. Toulmin points out the remarkable similarity between Europe from 1914 to 1945 and Europe during the Thirty Years War (1618–1648). In the face of the catastrophe in Central Europe which this brought – the end of the six hundred years old Habsburg Empire – the search for what is universal and certain in the manner of mathematics was pursued 'with *even greater* enthusiasm, and in an *even more extreme* form, than had been the case in the mid-seventeenth century' (*Cosmopolis*, p.159, author's italics; see pp.152–160). The difference is that, in the twentieth century, the paradigm was provided by the mathematical logic of Russell and Whitehead's *Principia Mathematica*, instead of the geometry of Euclid's *Elements*. The outcome this time, as a consequence of the mathematisation of logic, was a *universal* machine: the computer. This is something with which we still have to come to terms.

11. John Barrow emphasises that Newton's work led to more than a revolution in scientific thinking:

> It changed the thinking of non-scientists as well. The *Principia* became the first scientific 'cult' book (that is, a book that is read about, but not read), and it created what might be called 'Newtonianism'. This had many consequences, the most interesting of which was the start of the systematic popularisation of science through the publication of elementary explanations designed for the lay-person. A vast number of such books were written in the first half of the eighteenth century to satisfy public interest in Newton and his discoveries. (Barrow, *The World Within the World*, p.70.).

12. Clifford Geertz, quoted in Isaiah Berlin, *The Crooked Timber of Humanity: Chapters in the History of Ideas* (London: John Murray, 1990), p.70.

13. A Chinese physicist can help to remind us what a remarkable step it is. Heinz Pagels records that:

> Many years ago I asked T.D. Lee, a Nobel Laureate in physics
> born in China, about his educational experiences before he went
> to Chicago to study with the physicist Enrico Fermi. What had
> impressed him as a student in China when he first encountered
> physics? Without hesitation Lee replied that it was the idea that
> physical laws applied here on earth, in one's living room as well as
> on Mars, that was new and compelling to him. (Heinz R. Pagels, *The
> Cosmic Code: Quantum Physics as the Language of Nature*, London:
> Penguin Books, 1984, p.304.)

He goes on to comment:

> The universality of physical laws is perhaps their deepest feature ...
> This fact is rather surprising, for nothing is less evident in the variety
> of nature than the existence of universal laws.

The reason why this would impress Lee so strongly is possibly because such an idea of universality is not emphasised in Chinese culture. Instead, priority is given to the uniqueness of the particular case, seen in the context of other such particulars, instead of looking for an underlying unity. So instead of the notion of universal law in the Western sense, in the Chinese conception everything has its own law within it according to its nature. This example enables us to see that nature can manifest in different aspects according to differences in the cultural context, so that what seems 'obvious' to us may be only one possibility. It shows us that our ideal of universality is not culturally universal.

14. Lawrence Schmidt, "Uncovering Hermeneutic Truth", in Schmidt, ed. *The Specter of Relativism*, p.75.

15. *Ibid.* p.76.

16. A very clear illustration of this is given in the case of the philosophy of religion in Jorge N. Ferrer, *Revisioning Transpersonal Theory: A Participatory Vision of Human Spirituality* (Albany: State University of New York Press, 2002). Ferrer contrasts the universalism of the perennial philosophy – which asserts a single Truth underlying the multiplicity of religious traditions – with the contrary view of contextualism, which asserts the pure plurality of these traditions. Ferrer shows that the way to get off the see-saw of objectivism versus relativism to which this leads, can be found by drawing on the resources of epistemology in analytical philosophy. After identifying the Cartesian presuppositions which are usually unnoticed underlying perennialism, and the Neo-Kantian roots of contextualism, he goes on to

show that these epistemological positions subscribe to the Myth of the Given (Sellars), the Myth of the Framework (Popper) , and the Dualism of Framework and Reality (Davidson) – all of which have been undermined by contemporary epistemology:

> Both approaches, however are burdened by a host of Cartesian-Kantian prejudices that not only reduce their explanatory power, but also force spiritual possibilities into very limited moulds. More specifically, perennialism and contextualism are both contingent on the Dualism of Framework and Reality (i.e., a vision of human knowledge as mediated through conceptual frameworks which can neither directly access nor fully convey a supposedly uninterpreted reality). This basic dualism naturally engenders two interdependent epistemological myths: The Myth of the Given (there is a single pre-given reality out there independent of any cognitive activity), and the Myth of the Framework (we are epistemological prisoners trapped in our conceptual frameworks). Although representatives of these approaches tend to subscribe to both myths to some degree, perennialists tend to be particularly bewitched by the Myth of the Given, while contextualists tend to be especially constrained by the Myth of the Framework. These epistemological myths, we have seen here, not only create all sorts of pseudo-problems about the nature of spiritual knowing, but also contribute in fundamental ways to human alienation by severing our direct connection with the source of our being. (p.156)

Once this work of deconstruction is done, Ferrer shows that the way is open to understanding religious insight as a participatory event – as 'an ontological "happening" of Being in the locus of human historical existence' (p.118). He refers explicitly to Gadamer's notion of truth as an event of disclosure, quoting Gadamer's words: 'Being is self-presentation and ... all understanding is an event' (*Ibid.;* Gadamer, *Truth and Method*, p.484). Once the Myth of the Given and the Dualism of Framework and Reality are disposed of, we can see that:

> In a participatory epistemology free from these Cartesian-Kantian moulds, the so-called mediating principles (languages, symbols, etc.) are no longer imprisoning, contaminating, or alienating barriers that prevent us from a direct, intimate contact with the world. On the contrary, once we accept that there is not a pregiven reality to be mediated, these factors are revealed as the vehicles through which reality or being self-manifests in the locus of the human.

> Like Gadamer's revision of the nature of historical prejudices, that
> is, *mediation is transformed from being an obstacle into the very means*
> *that enables us to directly participate in the self-disclosure of the world.*
> (p.172)

Ferrer's work illustrates brilliantly, and in turn I suggest is illuminated by, the dynamic understanding of being in which difference is ontological.

17. Zimmerman, *Eclipse of the Self,* p.1.
18. Thomas Sheehan, 'Dasein', in Dreyfus and Wrathall, eds., *A Companion to Heidegger,* p.206. Sheehan is referring to Aristotle, *De Anima,* Book 3, Chapter 7, 432a2.
19. *Anima est quodammodo omnia;* Aristotle, *De Anima,* book 3, Chapter 8 (431b20). This Latin translation is the one quoted by Aquinas in his *Summa Theologica* in the Middle Ages. Aquinas expressed Aristotle's non-dual philosophy of participation in his own way as *'intellectus in actu est intelligibile in actu',* which Kerr translates as 'our intellectual capacities actualised *are* the world's intelligibility realized'. See Fergus Kerr, *After Aquinas: Versions of Thomism* (Oxford: Blackwell, 2002), p.27.
20. Sokolowski, *Introduction to Phenomenology,* p.4.
21. *Ibid.* p.185.
22. Quoted in Chapter 2, note 59; McGilchrist, *The Master and his Emissary,* p.179.
23. Hans-Georg Gadamer, *The Relevance of the Beautiful and Other Essays* (Cambridge: Cambridge University Press, 1986), p.79. Gadamer goes on to say that 'what the gesture reveals is the being of meaning rather than the knowledge of meaning'. This expresses succinctly the difference between the lived experience (right brain) and the representation of experience (left brain).
24. Ludwig Wittgenstein, *Zettel* (Berkeley: University of California Press, 2012), section 225.
25. Ray Monk, *Ludwig Wittgenstein: The Duty of Genius* (London: Vintage, 1991), p.548. Monk tells an illuminating story about Wittgenstein in this regard:

> Once, when Wittgenstein and Drury were walking together in the
> west of Ireland, they came across a five-year-old girl sitting outside
> a cottage. 'Drury, just look at the expression on that child's face',
> Wittgenstein implored, adding: 'You don't take enough notice of
> people's faces; it is a fault you ought to try to correct.' *(Ibid.)*

26. Wittgenstein, *Zettel*, section 220.

27. Anthony Rudd, *Expressing the World: Skepticism, Wittgenstein, and Heidegger* (Chicago: Open Court, 2003), p.178. Rudd considers several candidates for perceiving the world expressively in a way that is similar to Wittgenstein's expressivist understanding of other minds. These include Romanticism, and the phenomenology of Heidegger and Wittgenstein.

28. Monk, *Ibid*. Monk draws attention to the influence which Goethe had on the later Wittgenstein. His practice of looking for a synoptic or perspicuous presentation, as he called it, came from his discovery of Goethe's morphological approach (Monk, pp.509–12). Wittgenstein first learned of this through his reading of Spengler's *Decline of the West*. Spengler follows what he calls a physiognomic method in the study of history, which he says was inspired by Goethe's notion of a morphological study of nature (Monk, p.303). Finch calls this physiognomic way of seeing, Wittgenstein's (and Goethe's) 'physiognomic phenomenalism'. See Henry Le Roy Finch, *Wittgenstein* (Shaftesbury: Element, 1995), p.61, and also Henry Le Roy Finch, *Wittgenstein – The Later Philosophy: An Exposition of the 'Philosophical Investigations'* (Atlantic Highlands: Humanities Press, 1977), pp.172–77. It is interesting that we can now reverse the influence and find Wittgenstein helpful in understanding Goethe.

29. Emma Kidd, 'Turning a New Leaf', *Holistic Science Journal*, Vol. 1, No. 2, November 2010.

Bibliography

Abley, Mark (2003) *Spoken Here: Travels Among Threatened Languages,* Heinemann, London.

Agamben, Giorgio (1999) *Potentialities: Collected Essays in Philosophy,* Stanford University Press, Stanford.

Amrine, Frederick, Zucker, Francis J., & Wheeler, Harvey, eds. (1987) *Goethe and the Sciences: A Reappraisal,* University of Chicago Press, Chicago.

Barrow, John D. (1988) *The World Within the World: A Journey to the Edge of Space and Time,* Oxford University Press, Oxford.

Beistegui, Miguel de, (2005) *The New Heidegger,* Continuum, London.

Bennett, J.G. (1956) *The Dramatic Universe, Volume 1: The Foundations of Natural Philosophy,* Hodder and Stoughton, London.

Berlin, Isaiah (1990) *The Crooked Timber of Humanity: Chapters in the History of Ideas,* John Murray, London.

Bernstein, Richard J. (1983) *Beyond Objectivism and Relativism: Science, Hermeneutics, and Praxis,* Blackwell, Oxford.

Bertalanffy, Ludwig von (1968) *General System Theory,* Braziller, New York.

Blanché, Robert (1968) *Contemporary Science and Rationalism,* Oliver and Boyd, Edinburgh.

Bohm, David (1980) *Wholeness and the Implicate Order,* Routledge and Kegan Paul, London.

Bortoft, Henri (1986) *Goethe's Scientific Consciousness,* Institute for Cultural Research Monograph Series No 22.

—, (1996) *The Wholeness of Nature: Goethe's Way of Science,* Floris Books, Edinburgh.

Bowler, Peter J. (1989) *Evolution: The History of an Idea,* University of California Press, Berkeley.

Brown, Harold I. (1977) *Perception, Theory and Commitment: The New Philosophy of Science,* University of Chicago Press, Chicago.

Bruns, Gerald L. (1992) *Hermeneutics Ancient and Modern,* Yale University Press, New Haven.

Burtt, E.A.(1932) *The Metaphysical Foundations of Modern Science,* Routledge and Kegan Paul, London.

Caird, Edward (2002) *Hegel,* Cambridge Scholars Press, London.

Cassirer, Ernst (1953) *Substance and Function and Einstein's Theory of Relativity,* Dover Publications, New York.

Cerbone, David R. (2006) *Understanding Phenomenology,* Acumen, Chesham.

Clark, Timothy (2002) *Martin Heidegger,* Routledge, London.

Cohen, I. Bernard (1993) *The Newtonian Revolution: with illustrations of the transformation of scientific ideas,* Cambridge University Press, Cambridge.

Cooper, David E. (1990) *Existentialism: A Reconstruction,* Blackwell, Oxford.

Courant, Richard, & Robbins, Herbert (1996) *What is Mathematics?: An Elementary Approach to Ideas and Methods,* Oxford University Press, Oxford.

Critchley, Simon (2001) *Continental Philosophy: A Very Short Introduction,* Oxford University Press, Oxford.

Crombie, A.C. (1953) *Robert Grosseteste and the Origins of Experimental Science 1100–1700,* Oxford University Press, Oxford.

Davis, Bret W., ed. (2010) *Martin Heidegger: Key Concepts,* Acumen, Durham.

Deleuze, Gilles (1991) *Bergsonism,* Zone Books, New York.

Desmond, Adrian, & Moore, James (1992) *Darwin,* Penguin Books, London.

Desmond, Adrian (1992) *The Politics of Evolution: Medicine and Reform in Radical London,* University of Chicago Press, Chicago.

Dillon, M.C. (1997) *Merleau-Ponty's Ontology,* Northwestern University Press, Evanston.

Dostal, Robert J., ed. (2002) *The Cambridge Companion to Gadamer,* Cambridge University Press, Cambridge.

Drake, Stillman (1957) *Discoveries and Opinions of Galileo,* Doubleday, New York.

Dreyfus, Hubert L. (1991) *Being-in-the-World: A Commentary on Heidegger's 'Being and Time', Division 1,* MIT Press, Cambridge MA.

—, & Wrathall, Mark A., eds. (2007) *A Companion to Heidegger,* Blackwell, Oxford.

Dunbar, Robin (1995) *The Trouble with Science,* Harvard University Press, Cambridge MA.

Eagleton, Terry (2007) *How to Read a Poem,* Blackwell, Oxford.

Easlea, Brian (1980) *Witch Hunting, Magic and the New Philosophy: An Introduction to the Debates of the Scientific Revolution,* Humanities Press, New Jersey.

Elliot, Jason (1999) *An Unexpected Light: Travels in Afghanistan,* Picador, London.

Ferrer, Jorge N. (2002) *Revisioning Transpersonal Theory: A Participatory Vision of Human Spirituality,* State University of New York Press, Albany.

Figal, Günter (2009) 'Hermeneutics as Phenomenology', *The Journal of the British Society for Phenomenology,* Volume Forty, Number Three.

Finch, Henry Le Roy (1977) *Wittgenstein – The Later Philosophy; An Exposition of the 'Philosophical Investigations',* Humanities Press, Atlantic Highlands.

Finch, Henry Le Roy (1995) *Wittgenstein,* Element, Shaftesbury.

Fink, Karl J. (1991) *Goethe's History of Science,* Cambridge University Press, Cambridge.

Gadamer, Hans-Georg (1976) *Hegel's Dialectic: Five Hermeneutic Studies,* Yale University Press, New Haven.

—, (1977) *Philosophical Hermeneutics,* University of California Press, Berkeley.

—, (1986) *The Idea of the Good in Platonic-Aristotelian Philosophy,* Yale University Press, New Haven.

—, (1986) *The Relevance of the Beautiful and Other Essays,* Cambridge University Press, Cambridge.

—, (1989) *Truth and Method,* Sheed and Ward, London.

Galileo (1967) *Dialogues Concerning the Two Chief World Systems,* University of California Press, Berkeley.

Goethe, Johann Wolfgang von (2009) *The Metamorphosis of Plants,* Introduction and Photography by Gordon L. Miller, MIT Press, Cambridge MA.

Goodwin, Brian (1994) *How the Leopard Changed its Spots: the Evolution of Complexity,* Weidenfeld and Nicolson, London.

Groarke, Louis (2009) *An Aristotelian Account of Induction: Creating Something from Nothing,* McGill-Queens University Press, Montreal.

Grohmann, Gerbert (1974) *The Plant: A Guide to Understanding its Nature,* Rudolf Steiner Press, London.

Grondin, Jean (1995) *Sources of Hermeneutics*, State University of New York Press, Albany.

Guignon, Charles, B. (1983) *Heidegger and the Problem of Knowledge*, Hackett, Indianapolis.

Gurwitsch, Aron (1974) *Phenomenology and the Theory of Science*, Northwestern University Press, Evanston.

Hamblyn, Richard (2001) *The Invention of Clouds: How an Amateur Meteorologist Forged the Language of the Skies*, Picador, London.

Harrison, K. David (2007) *When Languages Die: The Extinction of the World's Languages and the Erosion of Human Knowledge*, Oxford University Press, Oxford.

Heidegger, Martin (1962) *Kant and the Problem of Metaphysics*, Indiana University Press, Bloomington.

—, (1962) *Being and Time*, trans. John Macquarrie and Edward Robinson, SCM Press, London.

—, (1969) *Identity and Difference*, Harper and Row, New York.

—, (1971) *On the Way to Language*, Harper and Row, New York.

—, (1985) *History of the Concept of Time: Prolegomena*, Indiana University Press, Bloomington.

—, (2000) *Introduction to Metaphysics*, Yale University Press, New Haven.

—, (2001) *Zollikon Seminars: Protocols – Conversations – Letters*, Northwestern University Press, Evanston.

—, (2002) *Towards the Definition of Philosophy*, Continuum, London.

Hirsch, E.D., Jr. (1967) *Validity in Interpretation*, Yale University Press, New Haven.

Holdrege, Craig (1996) *Genetics and the Manipulation of Life: The Forgotten Factor of Context*, Lindisfarne Press, Hudson.

Hübner, Kurt (1983) *Critique of Scientific Reason*, University of Chicago Press, Chicago.

Husserl, Edmund (1964) *The Idea of Phenomenology*, Martinus Nijhoff, The Hague.

Humbolt, Wilhelm von (1999) *On Language: On the Diversity of Human Language Construction and Its Influence on the Mental Development of the Human Species*, Cambridge University Press, Cambridge.

Janik, Allan, and Toulmin, Stephen (1973) *Wittgenstein's Vienna*, Simon and Schuster, New York.

Johnson, Mark (1987) *The Body in the Mind: The Bodily Basis of Meaning, Imagination, and Reason*, University of Chicago Press, Chicago.

Johnson, Patricia Altenbernd (2000) *On Gadamer*, Wadsworth, Belmont.

Joseph, George Gheverghese (2000) *The Crest of the Peacock: The Non-European Roots of Mathematics*, Penguin Books, London.

Keller, Helen (1959) *The Story of My Life*, Hodder and Stoughton, London.

Kerr, Fergus (2002) *After Aquinas: Versions of Thomism*, Blackwell, Oxford.

Kline, Morris (1980) *Mathematics: The Loss of Certainty*, Oxford University Press, New York.

Koestler, Arthur (1959) *The Sleepwalkers: A History of Man's Changing Vision of the Universe*, Penguin Books, London.

Koyré, Alexandre (1957) *From the Closed World to the Infinite Universe*, John Hopkins University Press, Baltimore.

Kühlewind, Georg (1991) *The Logos-Structure of the World: Language as Model of Reality*, Lindisfarne Press, Hudson.

Kuhn, Thomas S. (1957) *The Copernican Revolution: Planetary Astronomy in the Development of Western Thought*, Harvard University Press, Cambridge MA.

Lear, Jonathan (1988) *Aristotle: the desire to understand*, Cambridge University Press, Cambridge.

Losee, John (1993) *A Historical Introduction to the Philosophy of Science*, Oxford University Press, Oxford.

Madison, G.B. (1988) *The Hermeneutics of Postmodernity: Figures and Themes*, Indiana University Press, Bloomington.

Magee, Bryan (1997) *Confessions of a Philosopher: A Journey Through Western Philosophy,* Weidenfeld and Nicolson, London.

May, Todd (2005) *Gilles Deleuze: An Introduction,* Cambridge University Press, Cambridge.

McGilchrist, Iain (2009) *The Master and His Emissary: The Divided Brain and the Making of the Western World,* Yale University Press, New Haven.

Merleau-Ponty, Maurice (1962) *Phenomenology of Perception,* Routledge and Kegan Paul, London.

—, (1973) *Consciousness and the Acquisition of Language,* Northwestern University Press, Evanston.

—, (1973) *The Prose of the World,* Northwestern University Press, Evanston.

Miller, Douglas, ed. (1988) *Goethe: Scientific Studies,* Suhrkamp, New York.

Mitchell, David (1962) *An Introduction to Logic,* Hutchinson, London.

Monk, Ray, (1991), *Ludwig Wittgenstein: The Duty of Genius,* Vintage, London.

Ornstein, Robert E., ed. (1973) *The Nature of Human Consciousness,* W.H. Freeman, San Francisco.

Pagels, Heinz R. (1984) *The Cosmic Code: Quantum Physics as the Language of Nature,* Penguin Books, London.

Palmer, Richard E. (1969) *Hermeneutics: Interpretation Theory in Schleiermacher, Dilthey, Heidegger, and Gadamer,* Northwestern University Press, Evanston.

—, (1973) 'Phenomenology as Foundation for a Post-Modern Philosophy of Literary Interpretation', *Cultural Hermeneutics,* Vol. 1.

Peat, F. David (1996) *Blackfoot Physics: A Journey into the Native American Universe,* Fourth Estate, London.

Proskauer, Heinrich, O. (1986) *The Rediscovery of Colour: Goethe versus Newton Today,* Anthroposophic Press, Spring Valley.

Pylkkänen, Paavo, T.I. (2007) *Mind, Matter and the Implicate Order,* Springer, Berlin.

Richards, Robert J. (2002) *The Romantic Conception of Life: Science and Philosophy in the Age of Goethe,* University of Chicago Press, Chicago.

Risser, James (1997) *Hermeneutics and the Voice of the Other: Re-reading Gadamer's Philosophical Hermeneutics,* State University of New York Press, Albany.

Rudd, Anthony (2003) *Expressing the World: Skepticism, Wittgenstein, and Heidegger,* Open Court, Chicago.

Sacks, Oliver (1986) *The Man Who Mistook His Wife For A Hat,* Pan Books, London.

Safranski, Rüdiger (1998) *Martin Heidegger: Between Good and Evil,* Harvard University Press, Cambridge MA.

Scheibler, Ingrid (2000) *Gadamer: Between Heidegger and Habermas,* Rowman and Littlefield, Lanham MA.

Schelling, F.W.J. (2004) *First Outline of a System of the Philosophy of Nature,* State University of New York Press, Albany.

Schmidt, Lawrence K., ed. (1995) *The Specter of Relativism: Truth, Dialogue, and 'Phronesis' in Philosophical Hermeneutics,* Northwestern University Press, Evanston.

Scott, Charles E. (2002) *The Lives of Things,* Indiana University Press, Bloomington.

Seamon, David, & Zajonc, Arthur (1998) *Goethe's Way of Science: A Phenomenology of Nature,* State University of New York Press, Albany.

Sepper, Dennis L. (1988) *Goethe Contra Newton: Polemics and the project for the new science of colour,* Cambridge University Press, Cambridge.

Seymour, John (1977) *The Countryside Explained,* Faber & Faber, London.

Sokolowski, Robert (2000) *Introduction to Phenomenology,* Cambridge University Press, Cambridge.

Sorell, Tom (1987) *Descartes,* Oxford University Press, Oxford.

Stefanovic, Ingrid Lemon (2000) *Safeguarding Our Common Future: Rethinking Sustainable Development,* State University of New York Press, Albany.

Steiner, Rudolf (1968) *A Theory of Knowledge Based on Goethe's World Conception,* Anthroposophic Press, New York.

—, (1985) *Goethe's World View,* Mercury Press, Spring Valley.

—, (2000) *Nature's Open Secret: Introduction to Goethe's Scientific Writings,* Anthroposophic Press, Great Barrington.

Taylor, Charles (1989) *Sources of the Self: The Making of the Modern Identity,* Cambridge University Press, Cambridge.

Toadvine, Ted (2009) *Merleau-Ponty's Philosophy of Nature,* Northwestern University Press, Evanston.

Toulmin, Stephen (1957) *Foresight and Understanding: An enquiry into the aims of Science,* Hutchinson, London.

—, (1992) *Cosmopolis: The Hidden Agenda of Modernity,* University of Chicago Press, Chicago.

—, (2001) *Return to Reason,* Harvard University Press, Cambridge MA.

—, (2003) *The Uses of Argument,* Cambridge University Press, Cambridge

Wachterhauser, Brice R. (1999) *Beyond Being: Gadamer's Post-Platonic Hermeneutic Ontology,* Northwestern University Press, Evanston.

—, ed. (1986) *Hermeneutics and Modern Philosophy,* State University of New York Press, Albany.

—, ed. (1994) *Hermeneutics and Truth,* Northwestern University Press, Evanston.

Watts, Michael (2011) *The Philosophy of Heidegger,* Acumen, Durham.

Weinsheimer, Joel C. (1985) *Gadamer's Hermeneutics: A Reading of 'Truth and Method',* Yale University Press, New Haven.

—, (2002) *Philosophical Hermeneutics and Literary Theory,* Yale University Press, New Haven.

Wertheim, Margaret (1997) *Pythagoras' Trousers: God, Physics and the Gender Wars,* Fourth Estate, London.

Westfall, Richard S. (1977) *The Construction of Modern Science: Mechanisms and Mechanics,* Cambridge University Press, Cambridge.

Whitehead, Alfred North (1979) *Process and Reality,* Free Press, New York.

Wittgenstein, Ludwig (2012) *Zettel,* University of California Press, Berkeley.

Zahavi, Dan (2003) *Husserl's Phenomenology,* Stanford University Press, Stanford.

Zimmerman, Michael E. (1986) *Eclipse of the Self: The Development Heidegger's Concept of Authenticity,* Ohio University Press, Athens.

—, (1990) Heidegger's Confrontation with Modernity: Technology, Politics, Art, Indiana University Press, Bloomington.

Index

233

Related books

THE WHOLENESS OF NATURE

GOETHE'S WAY OF SCIENCE

Henri Bortoft

Examines the phenomenological and cultural roots of Goethe's ways of science and argues that Goethe's insights represent the foundation for a future science.

www.florisbooks.co.uk

GOETHE ON SCIENCE

AN ANTHOLOGY OF GOETHE'S SCIENTIFIC WRITINGS

Edited by Jeremy Naydler

GOETHE
ON SCIENCE

An Anthology of
Goethe's Scientific Writings

Selected and introduced by Jeremy Naydler
With a foreword by Henri Bortoft

A systematic arrangement of extracts from Goethe's major scientific works which reveal how fundamentally different his approach was to scientific study of the natural world.

www.florisbooks.co.uk

NATURE'S DUE

HEALING OUR
FRAGMENTED CULTURE

Brian Goodwin

Challenges modern ideas on the interaction of science, nature and human culture, with far-reaching consequences for how we govern our world.

www.florisbooks.co.uk